CALGARY PUBLIC LIBRARY

OCT 2009

"The mere facts that we see things, that [vision is] a sort of remote sensing, and that it is brought about mostly subconsciously but that we experience the outcome consciously—all these core facts of perception are fascinating and often beyond our imagination. As his long-term mentor/collaborator, I can guarantee that Mark Changizi is 'a pencil-neck square, stick-in-the-mud scientist' who is most imaginative, creative, and entertaining in his writing. This book will no doubt offer a revolutionary view on our daily experience of visual perception. A superb reading for young students, scientists, businessmen, and others desperately seeking signs of incredible mental powers in themselves."

—Shinsuke Shimojo, Professor of Biology/Computation and Neural Systems, California Institute of Technology

"Changizi's pioneering research shows that evolutionary theory can help explain many of the great puzzles of vision. In this remarkable book, he fleshes out his findings and provides a fresh take on many key issues in perception. Psychologists and neuroscientists will be stimulated to incorporate evolutionary theory in their own research; non-specialists will appreciate the accessible, entertaining style. Highly recommended for all who are interested in evolution or perception."

—Robert Deaner, Assistant Professor of Psychology, Grand Valley State University

"Changizi is bristling with ideas. Rarely will one come across someone this capable of creative thinking. His ideas are, without doubt, speculative, and await further experimental confirmation. However, were that not true, his influence on visual neuroscience would be such that a majority of present-day neuroscientists would be working on ideas forwarded by him. Changizi has the unique ability to draw the reader into asking the most fundamental questions of 'why' rather than the more mundane ones of 'how' regarding the workings of the visual nervous system."

—Romi Nijhawan, Reader in Psychology, University of Sussex

"Most of us take for granted that we can open our eyes and see, even though how we do this is remarkably complex and still a mystery to science. Changizi is someone who has thought deeply about this problem—by asking not only 'how' we see, but 'why.' The result is an enthralling and insightful journey into the human mind, with much of the terrain revealed by the author's own discoveries. This is a book that will open your eyes to the amazing feats of your visual system."

—MICHAEL A. WEBSTER, Foundation Professor of Psychology, University of Nevada, Reno

"Mark Changizi has written a book full of invention and originality. It contains a quite remarkable number of ingenious observations, insights, and basic theory about the visual system. If you want to learn how to think outside of the box, then this is a book for you."

—PETER LUCAS, Professor of Anthropology, George Washington University

"The book was both fascinating and entertaining—one of the most original accounts of vision that I have ever read. Many of the ideas are speculative, of course—but this is precisely the strength of the book: to present entirely novel ideas that are sure to radically change your mind about the way vision works."

—STANISLAS DEHAENE, Director of the INSERM-CEA Cognitive Neuroimaging Laboratory, Chair of Experimental Cognitive Psychology, Collège de France

"It is written in a way that draws the reader in, even those who may not be too interested in vision. [Changizi] ask[s] the basic questions, ones so basic that many of us have not really thought about them, and then seeks to answer them. The book is nicely illustrated, raises and answers many questions about vision, and has an evolutionary perspective."

—JON H. KAAS, Centennial Professor of Psychology, Vanderbilt University

THE VISION REVOLUTION

THE VISION REVOLUTION

How the Latest Research Overturns Everything We Thought We Knew About Human Vision

Mark Changizi

BENBELLA BOOKS, INC.
DALLAS, TEXAS

Copyright © 2009 by Mark Changizi
Illustrations © 2009 by Ralph Voltz

All rights reserved. No part of this book may be used or reproduced in any manner whatsoever without written permission except in the case of brief quotations embodied in critical articles or reviews.

BenBella Books, Inc.
6440 N. Central Expressway, Suite 503
Dallas, TX 75206
www.benbellabooks.com
Send feedback to feedback@benbellabooks.com

Printed in the United States of America
10 9 8 7 6 5 4 3 2 1

Library of Congress Cataloging-in-Publication Data is available for this title.
ISBN 978-1933771-66-3

Proofreading by Emily Chauviere and Stacia Seaman
Cover design by Laura Watkins
Cover illustration by Mondolithic Studios Inc.
Text design and composition by John Reinhardt Book Design
Printed by Bang Printing

Distributed by Perseus Distribution
perseusdistribution.com

To place orders through Perseus Distribution:
Tel: 800-343-4499
Fax: 800-351-5073
E-mail: orderentry@perseusbooks.com

Significant discounts for bulk sales are available.
Please contact Glenn Yeffeth at glenn@benbellabooks.com or (214) 750-3628.

Contents

INTRODUCTION
Super Naturally

IN THE MOVIE *Unbreakable* by M. Night Shyamalan, the villain Elijah Price says, "It's hard for many people to believe that there are extraordinary things inside themselves, as well as others." Indeed, the story's superhero, David Dunn, is unaware of his super strength, his inability to be injured (except by drowning), and his ability to sense evil. Dunn would have lived his life without anyone—including himself—realizing he had superpowers if *Unbreakable*'s villain hadn't forced him into the discovery.

At first glance we are surprised that Dunn could be so in the dark about his abilities. How could he utilize his evil-detection power every day at work as a security guard without realizing he had it? However, aren't most powers—super or otherwise—like that? For example, our ability to simply stand requires complex computations about which we are unaware. Complex machines like David Dunn and ourselves only function because we have a tremendous number of "powers" working in concert, but we can only be conscious of a few of these powers at a time. Natural selection has seen to it that precious consciousness is devoted where it's most needed—and least harmful—leaving everything else running unnoticed just under the surface.

The involuntary functions of our bodies rarely announce their specific purposes. Livers never told anyone they're for detoxification, and they don't come with user's manuals. Neurosurgeons have yet to find any piece of brain with a label reading, "Crucial for future-seeing. Do not remove without medical or clerical consultation." The functions of our body are carried out by unlabeled meat, and no gadget—no matter how fancy—can allow us to simply read off those functions in a lab.

Powers are even harder to pin down, however, because they typically work superbly only when we're using them where and when we're supposed to. Our abilities evolved over millions of years to help us survive and reproduce in nature, and so you can't understand them without understanding the environment they evolved for, any more than you can understand a stapler without knowing what paper is.

Superpowers, then, can't be introspected. They can't be seen with a microscope. And they can't be grasped simply by knowing the ins and outs of the meat. Instead, the natural environment is half the story. Lucky for us there are ways of finding our powers. Science lets us generate a hypothesis concerning the purpose of some biological structure—what its power is—and then test that hypothesis and its predictions. Those predictions might concern how the power would vary with habitat, what other characteristics an animal with that power would be expected to have, or even what that biological structure would look like were it really designed with that power in mind. That's how we scientists identify structures' powers.

And that's what *this* scientist is doing in this book: identifying powers. Specifically, *super*powers. Even more specifically, superpowers of *vision*—*four* of them, one from each of the main subdisciplines of vision: color, binocularity, motion, and object recognition. Or in superhero terms: *telepathy*, *X-ray vision*, *future-seeing*, and *spirit-reading*. Now, you might be thinking, "How could we possibly have such powers? Mustn't this author be crazy to suggest such a thing?" Let me immediately allay your fears: there's nothing spooky going on in this book. I'm claiming we have these four superpowers, yes, but also that they are carried out by our real bodies and brains, with no mysterious mechanisms, no magic, and no funny business. Trust me—I'm a square, stick-in-the-mud, pencil-necked scientist who gets annoyed when one of the cable science channels puts a show on about "hauntings," "mystics," or other nonsense.

But then why am *I* writing about superpowers? "No magic, no superpowers," some might say. Well, perhaps. But I'm more inclined to say, "No magic, but *still* superpowers." I call each of these four powers "superpowers" because each of them has been attributed to superhuman characters, and each of them has been presumed to be well beyond the limits of us regular folk.

That we have superpowers of vision—and yet no one has realized it—is one of the reasons I think you'll enjoy this book. Superpowers *are* fun, after all. There's no denying it. But superpowers are just a part of this book's story. Each of the four superpowers is the tip of an iceberg, and lying below the surface is a fundamental question concerning our nature. This book is really about answering "why": *Why do we see in color? Why do our eyes face forward? Why do we see illusions? Why are letters shaped the way they are?*

What on Earth is the connection between these four deep scientific questions and the four superpowers? I'd hate to give away all the answers now—that's what the rest of the book is for—but here are some teasers. We use color vision to see skin, so we can sense the emotions and states of our friends and enemies (*telepathy*). Our eyes face forward so that we can see through objects, whether our own noses or clutter in the world around us (*X-ray vision*). We see illusions because our brain is attempting to see the future in order to properly perceive the present (*future-seeing*). And, lastly, letters have culturally evolved over centuries into shapes that look like things in nature because nature is what we have evolved to be good at seeing. These letters then allow us to effortlessly read the thoughts of the living...and the dead (*spirit-reading*).

Although the stories behind these superpowers concern vision, they are more generally about the brain and its evolution. Half of your brain is specialized for performing the computations needed for visual perception, and so you can't study the brain without spending about half your energies on vision; you won't miss out on nearly as much by skipping over audition and olfaction. And not only is our brain "half visual," but our visual system is by far the most well-understood part of our brains. For a century, vision researchers in an area called visual psychophysics have been charting the relationship between the stimuli in front of the eye and the resultant perception elicited "behind" them, in the brain. For decades neuroanatomists such as John Allman, Jon Kaas, and David Van Essen have been mapping the visual areas of the primate brain, and countless other researchers have been characterizing the functional specializations and mechanisms within these areas.

Furthermore, understanding the "why" of the brain requires understanding our brain's evolution and the natural ecological conditions that prevailed during evolution, and these, too, are much better understood for vision than for our other senses and cognitive and behavioral attributes. Although about half the brain may be used for vision, much more than half of the best understood parts of the brain involve vision, making vision part and parcel of any worthwhile attempt to understand the brain.

And who am I, in addition to being a square, stick-in-the-mud, pencil-necked cable viewer? I'm a theoretical neuroscientist, meaning I use my training in physics and mathematics to put forth and test novel theories within neuroscience. But more specifically, I am interested in addressing the function and design of the brain, body, behaviors, and perceptions. What I find exciting about biology and neuroscience is why things are the way they are, not how they actually work. If you describe to me the brain mechanisms underlying our perception of color, I'll still be left with what I take to be the most important issue: *Why* did we evolve mechanisms that implement that kind of perception in the first place? That question gets at the *ultimate* reasons for why we are as we are, rather than the *proximate* mechanical reasons (which make my eyes glaze over). In attempting to answer such "why" questions I have also had to study evolution, for only by understanding it and the ecological conditions wherein the trait (e.g., color vision) evolved can one come to an ultimate answer. So I suppose that makes me an evolutionary theoretical neuroscientist. That's why this book is not only about four novel ideas in vision science, but puts an emphasis on the "evolution" in "r*evolution*."

But enough with the introductions. Let's get started. Or perhaps I should say... up, up, and away!

CHAPTER 1
Color Telepathy

"Ho! Ho!" said the Leopard. "Would it surprise you very much to know that you show up in this dark place like a mustard-plaster on a sack of coals?"

"Well, calling names won't catch dinner," said the Ethiopian. "The long and little of it is that we don't match our backgrounds. I'm going to take Baviaan's advice. He told me I ought to change; and as I've nothing to change except my skin I'm going to change that."

"What to?" said the Leopard, tremendously excited.

"To a nice working blackish-brownish colour, with a little purple in it, and touches of slaty-blue. It will be the very thing for hiding in hollows and behind trees."

So he changed his skin then and there, and the Leopard was more excited than ever; he had never seen a man change his skin before.

—Rudyard Kipling,
"How the Leopard Got His Spots" (*Just So Stories*)

Empaths

Standing over broken glass and spilled milk you find your two children, each pointing at the other. Mind-reading would be mighty useful right now—interrogations of four-year-olds can be fruitless, and you'd hate to punish both. Unfortunately, the superhuman characters who can mind-read are unlikely to come to your rescue. They're busy sensing the emotions of aliens on starships with readied torpedoes or saving humankind from genocidal mutants. You are on your own.

Might *you* have the ability to read minds or sense emotions? If by "sensing emotions" we just mean the ability to recognize facial expressions, then, sure, we can all do that. But that is not what is typically meant by mind-reading. We expect mind-readers to employ something extra, some specialized external sensing equipment capable of receiving emotion-related signals. When we read facial expressions, we are not using a sixth sense; we're just using our eyes, and then our brains, to compute the visual information. That's why recognizing facial expressions does not count as mind-reading.

Mind-reading—i.e., having specialized sensing equipment that lets us figure out what others are thinking—is not as crazy as it first sounds. There are animals with the ability to actually sense another animal's brain activity: platypuses, eels, and sharks have specific organs that give them the power of *electroreception,* which is used to sense the presence of nervous system electrical activity in nearby animals. Since these animals may not be able to discriminate between different kinds of brain activity, this may not qualify as mind-reading, but it does show that there are animals able to sense brain activity remotely. Eels, unfortunately, are unlikely to be able to tell you which child spilled the milk, and will more than likely just cause other problems in the kitchen.

Mind-reading is actually done today in certain neuroscience labs using technology such as functional magnetic resonance imaging (fMRI), which allows scientists to see brain activity. For example, neuroscientists Yukiyasu Kamitani and Frank Tong showed lines of varied orientation to people lying in an fMRI machine, and were able to determine from the brain images which of eight orientations the person was seeing. Your children probably were not lying in fMRI machines when the milk was spilled, but you could recruit a much cruder invention to get to the bottom of the mystery after the fact: a lie-detector test. Lie detectors, which were invented in the late nineteenth century and have been used by law enforcement since, amount to a primitive mind-reading device. They measure physiological changes in the body such as heart rate, breathing rate, skin conductance, and blood pressure, relying on the fact that certain emotions or moods are typically accompanied by physiological changes, something long studied in the field of psychophysiology.

When a person's physiological state changes, there are often changes in the distribution of blood to various parts of the body. This, in turn, modulates the color and other properties of skin. In 1972, a bioengineer named Takuo Aoyagi capitalized on this idea by inventing the pulse oximeter, which sensed the color properties of skin and used that information to determine how oxygenated the blood was beneath it and how the blood volume changed over time. These devices can be found today in nearly every hospital room—even your naughty children probably had one placed on their feet at birth—but are used for monitoring a patient's overall physiological condition rather than reading their emotions. However, because emotions and moods are known to modulate our physiological condition, the pulse oximeter is yet another potential mind-reading device.

So, is mind-reading purely the province of superheroes, platypuses, police officers, and doctors? Or might we all have the superhuman ability to read minds to some extent? As we will see, we have our own special equipment designed to sense blood beneath bare skin, and it serves as a window through which we can read the minds of others. In fact, we have the ability to sense the same two properties of blood that Takuo Aoyagi's pulse oximeter measures: its oxygenation and its concentration. What is this special sensing equipment of ours that I am referring to? The color receptors in our eyes and, more generally, our color vision. Our eyes measure the same two blood variables as the pulse oximeter, in a very similar manner—by sensing the color of the skin. Our color vision is, then, a type of ancient oximeter, and just as oximeters can, in principle, be used to sense moods and emotions, our color vision has given us the ability to mind-read like an empath. This chapter takes up this story of color and empaths, a discovery I first published with Qiong Zhang and Shinsuke Shimojo in 2006 in the *Proceedings of the Royal Society of London*. The story begins with our skin.

Naked Colors

Next to parrots, fish, chameleons, squid, bees, flowers, and fruit, we humans come across as an uncolorful species. The color of our skin is not likely to be used for a carnival ride or a box of cereal, and we are not likely candidates for whimsical garden decorations when competing against flamingos and insect whirligigs. "Skin" color, no matter your race, is not a popular paint color for interior design. In addition to being colorful, many of the animals people do employ in their gardens also have the ability to see in color, such as birds, fish, reptiles, and bees, suggesting that being colorful and seeing in color might well go together (although not always—colorful cuttlefish are color blind). Therefore, on the basis of our dull natural appearance and unsuitability for playful garden decor or flashy advertising, an alien surmising about the human race might guess we can't see in color.

But we *can* see in color. An immediate tip-off to an alien observer would be that although *we* may appear dull, our cultural artifacts do not. Our clothes are vibrantly colored, as are our faces; fashion-savvy women can easily spend nearly thirty minutes a day applying color to their face. And not only do we color our bodies, we color our homes as well, and are willing to argue heatedly with a spouse about whether our walls should be Lucent Yellow or Wheat Grass. We have opinions about the color of our toothbrushes, toasters, dish soap, trash cans, pens, computers, and even our toilets. One would be hard-pressed to find many human products that cannot be purchased in more than one color. Henry Ford saying that "Any customer can have a car painted any color that he wants so long as it is black" is so memorable *because* such a lack of color choice is so rare; it implicitly recognizes our human obsession with color.

The colors found in our culture are, then, a dead giveaway that we have color vision. But having a colorful culture is a somewhat recent phenomenon, going back only thousands of years. Although people in tribal societies did paint their bodies and decorate themselves with tattoos and animal remains, overall, their cultural environment was typically less colorful than ours. Furthermore, as indicated by the fact that our color vision is shared by many other primates, we have been seeing in color for tens of millions of years, well before we had any cultural artifacts to add colors to. Therefore, whatever color vision is for, it is not primarily meant to see the colors found in culture. Rather, we use colors in culture because they tap into what we have evolved to see.

Why did we evolve color vision? Recent evidence suggests the reason has to do with skin. But given our uninteresting skin tones, one might initially discount this; color vision must be for seeing colorful things, right? Agreement with this reasoning is why, for a century, the main hypothesis (by nineteenth-century researcher Grant Allen, and contemporary anthropologists such as John Mollon, Daniel Osorio, and Misha Vorobyev) for why we see in color is that it allows us to see fruit against the backdrop of leaves while foraging—fruit is colorful. And more recently, it has been suggested by anthropologists Peter W. Lucas and Nathaniel J. Dominy that color is designed to let us see young, edible leaves—such leaves also are colorful. Color vision is for seeing the colorful, and because skin is not colorful, color vision cannot possibly be for seeing skin. Case closed.

But if color is not for seeing skin, why are we so infatuated with decorating our bodies and faces with color? To illustrate the extent to which we humans like to color ourselves, I measured the colors found in 1,813 pieces of Western clothing from the middle ages to 1800, from a book by Auguste Racinet (*Racinet's Full-Color Pictorial History of Western Costume: With 92 Plates Showing Over 950 Authentic Costumes from the Middle Ages to 1800*). Red was the most common, occurring nearly 20 percent of the time, followed by blue, white, and green. Colors broadly within the range of natural skin tones occurred in 8 percent of the cases. The results are summarized in Figure 1 (see color insert), and as you can see, nearly all the colors used in clothing are not very similar to skin tones. For millions of years our body decorations were confined to the natural, boring hues of skin and fur, and yet as soon as we could artificially decorate ourselves, we heaped on the color! Thus, we are left with the mystery of why we like to see colorful clothes in place of less-than-colorful skin.

Unless... our skin is more colorful than I have let on. Skin is not static. In addition to its powers of thermoregulation, resilience, and water repellence, skin has astonishing, almost magical, color properties that are not widely understood or appreciated. Skin has the ability to become colorful. And not only can skin acquire color, but skin, no matter one's race, can achieve *any* hue (a feat I'll explain later). Understanding the color powers of skin is crucial to understanding that color vision is meant to see skin (although it is also potentially useful for finding fruit and leaves), and in particular is meant for sensing moods, emotions, and other physiological states.

Green Photons

Before entering headlong into the story of our skin, it may help to set ourselves straight on what exactly color is—and is not—because many of our layman intuitions about color are misguided. In particular, we tend to believe that color vision is about perceiving wavelengths. That's a mistake. Wavelengths are certainly important in understanding color. After all, everyone knows that only some wavelengths of light are visible to us, namely those from about 400 nanometers (nm) to 700nm. And everyone also knows that low wavelengths appear to us as violet or blue, then higher wavelengths as green, yellow, orange, and eventually red. We see this correspondence between perception and physics most clearly when we see a rainbow.

The rainbow, however, is a red herring. In our retina, there are three kinds of neuron that respond to light, called cones. Each of these cone types—S, M, and L—is sensitive to a range of wavelengths of light, and will tend to fire when light in that range hits it. S, M, and L cones are sensitive, respectively, to short, medium, and long wavelength light. These—along with rods, which allow us to see when there is little light, like at night—are the most fundamental visual sensors we have, and all our visual perceptions are built from them. If color vision were only about discriminating individual wavelengths, we would need just *two* types of cone in our eyes instead of three. With one cone broadly sensitive to short wavelengths and the other sensitive to long wavelengths, every single wavelength of light would cause a different relative activation in the two cones. Even color-blind individuals perceive different wavelengths to differ in color (though for them the rainbow varies in color from blue through gray to yellow, something we'll discuss in greater detail later in the chapter).

Color vision in animals evolved to sense objects, not photons, and real objects in the world typically reflect *all* wavelengths of light to our eye. What varies from object to object is *the quantity* of each wavelength reflected. The amount of each wavelength the eye receives from an object is called the object's reflectance spectrum. Imagine that each one-nanometer interval from 400nm to 700nm can vary independently of every other, and imagine that each of these intervals can have ten different amounts of light. There are, then, ten different amounts of light a spectrum can have at the first interval at 400nm, and for each of these there are ten different amounts of light a spectrum can have at 401nm. That makes $10^2 = 100$ possible ways light can come to your eye, *just* considering light of wavelengths in those two intervals. With 300 intervals, there are 10^{300} possible object reflectance spectra. That is, there are 10^{300} different kinds of surfaces—each with a different reflectance spectrum—that could send light to your eye. This number is very close to infinity.

Luckily, animals with color vision aren't interested in distinguishing between every possible surface that might exist. Instead, animals sample just certain spots in the spectrum, spots that allow them to distinguish between the kinds of surfaces they need to care about to survive. Our eyes are skimpy spectrometers; rather than having 300 different kinds of wavelength-sensitive cone, one for each nanometer interval, we have only three cones, sampling three different regions of the wavelength spectrum. But by comparing the activity from just these three cones, we can tell the difference between crucial surfaces. A consequence of getting by with three cones is that there are objects with different spectral reflectance distributions that look identical in color to us. Just as you can tell the difference between red and green socks but your color-blind friend cannot, a bird with four cone types can distinguish between colors that look identical to your human eyes. As we will see later in this chapter, our cone sensitivities are just right for distinguishing between the different spectral changes that occur in skin as a function of the underlying blood physiology.

The Special "Uncoloriness" of Your Skin

I'm looking around my family room at the moment, and I invite you to look around at your own current surroundings. What colors are the objects around you? My couch is reddish-brown, the walls white, the computer desktop blue, and a toy duck's beak orange; my books come in all kinds of colors. I have no problem naming any of them, and neither does my three-year-old daughter. And of course, you probably don't have any trouble naming the colors around you either.

But now ask yourself, what is the color of your own skin? The odd thing about skin color is that, in contrast to the ease with which we can name the colors of everything around us, no adequate color term seems to apply to our own skin color. While it's easier to say someone of a different color than you is "white," "pink," "brown," or "black," it is difficult to describe one's own skin (or the most common skin found in one's community, which over evolutionary history tended to be the same as your own skin). Caucasians might say that their skin is peach-colored, but they will admit it is not a good answer. Tan is not quite right either. Pink? No. There is simply no good word for it in English. Those of African ancestry might say that their skin is brown or chocolate-colored, but also admit that these are not quite right. Throughout history, the actual terms used in culture to refer to skin color have been determined by the dominant race. For example, Crayola crayons used to have a "skin" colored crayon, which they later changed to the more politically correct "peach."

The lack of a proper color term for referring to skin tone is not a problem unique to the English language. It appears to be a problem for all languages. Over the last fifty years, evidence has accumulated showing that people all see the same colors, and that languages worldwide tend to carve up the visual space of colors in the same way. This may seem obvious, but the academic community disagreed when anthropologists Brent Berlin and Paul Kay first proposed and defended the idea in the 1960s. In particular, one of their key discoveries was that all languages distinguish at most eleven different colors: white, gray, black, blue, green, yellow, orange, red, brown, pink, and purple. Most languages also have other colors in addition to these, but unlike the fundamental eleven mentioned above, the others are used differently, and their meaning is often unknown by a significant fraction of the population. For example, most people who speak English know these eleven colors, but are likely confused about the meaning of cyan, mauve, and chartreuse.

The relevance of these eleven basic colors to our discussion is that none of them apply well to skin. Furthermore, the vision scientist Robert Boynton has shown that skin colors are among the colors least served by these basic eleven. That is, most possible colors can be lumped under one of these eleven categories, even if it is a bit of a stretch. But the colors found in human skin tend to require the greatest stretching. For this range of colors, there is no consensus about what color term applies. For example, Robert Boynton found that, when eighteen people were asked to name the most uncategorizable color in the skin color range, five responded with "peach," four with "tan," three with "brown," three with "salmon," two used "orange," and one person "pink." This demonstrates that the color range for which languages have no adequate name coincides with the range in which human skin tones are found.

One's own skin color, then, tends to be difficult to name. But that's not all. Additionally, the experience of seeing skin color is not quite the same as the experience of seeing other colors. To illustrate this, let's do a little experiment: find a photograph of someone (like in Figure 2), and name all the colors you see in the picture. Because you are probably not a naïve experimental subject any longer—you know too much about where I'm going with this—you might want to try the experiment on a friend.

This simple exercise is motivated by an observation I made while watching a television advertisement for Target. The commercial showed people, dressed completely in red and white, against an entirely red and white background. What I noticed was that there appeared to be only two colors used in the ad, red and white. Nothing else. Yet there were also people in the advertisement, and their faces and hands were certainly not painted red or white. I would have noticed that! What I had almost failed to notice was that there were actually *three* colors in the advertisement, not just two: red, white, and *skin* color. The perceptual experience of skin color was altogether different than the experience of the other two colors: the skin did not register in my brain as a color at all. The red and white appeared painted onto my visual experience, but the skin color appeared more as a *lack* of paint. And even the color white, which is often not labeled as a color at all, was easy to see, define, and name in the Target ad. I suspect that when you or your friend counted the colors in the photograph, skin color may have been overlooked.

When I first noticed this uncategorizable nature of our own skin color, it struck me as utterly remarkable. Not only is our own skin color one of the few colors that we have such trouble naming, but it is all the more surprising given the importance of skin. We look at skin all day! In fact, looking at other people's faces is one of the main things we do as humans. Much of our success or failure in our lives depends on getting along with those in our social group, and so skin is not merely something we commonly see, but an object of supreme value in our lives. What a coincidence, I thought, that skin is one of the most important things in our lives, and yet one of the few things that we have such difficulty naming—and one of the few things that appears strangely uncolored (even compared to white!). My hunch was that this was probably *not* a coincidence at all.

But why? Why would we evolve to perceive our own skin color as uncategorizable and uncolored? How could this be a useful thing? Consider an object with a color that *is* highly categorizable—say an orange. If I place 100 oranges in front of you, there will actually be some variation in their colors, but you won't pay much attention to these differences. You will subconsciously lump together all the different hues into the same category: "orange." Ignoring differences is a central feature of categorization. To categorize is to stereotype. When a color is *un*categorizable, however, the opposite of stereotyping occurs. Rather than lumping together all the different colors, you appreciate all the little differences. Because our skin color cannot be categorized, we are better able to see very minor deviations in skin color, and therefore register minor changes in others' skin color as they occur. This made me wonder whether this is no accident. Could our color vision have evolved so for this precise purpose?

Recall, too, something I mentioned earlier: we perceive skin color differently than we perceive other colors, even white. Skin color seems qualitatively uncolored. What could the reason be for this? Consider an analogy with taste. What does your own saliva taste like? It does not taste like anything, really. And the same is true for the smell of your nose, of course—it does not smell like anything. Also, your skin does not seem to have much of a temperature at all. We are designed to perceive *changes* from normal baseline, and that is why in these cases we are designed to not have a strong perceptual feeling of taste, smell, or temperature. But notice that as soon as the chemical contents in your mouth begin to deviate, even by a tiny bit, you taste it. By not tasting baseline saliva, you are better able to sense the other tastes. And the same goes for smell and temperature. Have you ever thought about how remarkable it is that you can tell *with your hand* when someone has a fever? Not only can you sense it, but the person with a fever really does feel hot, even though they may only be one degree warmer than you! This amazing ability to notice changes in temperature relies in part on the fact that our normal, baseline temperature feels like nothing. We really notice the difference between perceiving *nothing* and perceiving *something*.

In this light, the apparent "uncoloriness" of your own skin is just like the lack of perceived taste, smell, or temperature of your own body. Your skin color, like these other senses, has been calibrated to zero by your color perception system, and this lets us more easily see color deviations away from zero, i.e., away from the baseline color. So just as we are designed to taste minute changes from the baseline taste of our own saliva (we can taste only a few molecules of salt), or feel even tiny changes in our baseline temperature, our difficulty in perceiving and categorizing skin color suggests that the reason we have color vision is to perceive skin color changes away from the baseline.

And why might *this* be useful? Probably because one's skin color can change depending on one's mood or overall state, and being able to sense these moods in others can be a strong advantage. Uncolorful, uncategorizable skin tones are just what we'd expect if color vision were intended for mind-reading through the window of skin.

Ethnic Illusions

If our skin color is so uncolored, why do we use color terms so often to refer to race? Races may not literally be white, black, brown, red, or yellow, but we wouldn't use these terms if we didn't perceive other races as having colored skin. So what is all this nonsense about skin being uncolored?

One must remember that it is only one's own skin that appears uncolored. I perceive *my* saliva as tasteless, but I might be able to taste yours. I don't smell my nose, but I might be able to smell yours. Similarly, my own skin may appear uncolored to me, but as a consequence of using senses designed to perceive changes around a baseline, even fairly small deviations from that baseline are perceived as qualitatively colored, just as a slightly warmer temperature is perceived as hot. An alien coming to visit us would find it utterly perplexing that a white person perceives a black person's skin to be so different from his own, and vice versa, when in fact, their spectra are practically identical (see Figure 3). But then again, this alien would also be surprised to learn that you perceive 100-degree skin as hot, even though 98.6 degrees and 100 degrees are practically the same.

Therefore, the fact that languages tend to use color terms to refer to other races is not at all mysterious. It is consistent with what would be expected if our color vision was designed to see color changes around baseline skin color. Your baseline skin color appears uncategorizable and uncolored, whereas skin colors that deviate even a little from baseline appear categorizably colorful.

Skin color is a lot like voice accents. What is the accent of your own voice? The answer is that you perceive it to have no accent, but you perceive people coming from other regions or countries as having accents. Of course, they believe that *you* are the one with the accent, not them. This is because we are designed to ably distinguish voices of people who have the same accent (or non-accent) as we do. We differentiate different people's voices as well as the inflections in the voice of a single individual. A consequence of this is that our own voice and those typical of our community are perceived as non-accented, and even fairly small deviations from this baseline accent are perceived as accented (e.g., country, urban, Boston, New York, English, Irish, German, and Latino accents). Because of this, we find it difficult to recognize people who have an accent by voice. We also find it more difficult to recognize the tone or emotional inflections of a speaker with an accent.

In our earlier discussion about your perception of your own skin color, I was implicitly assuming that you and your community share approximately the same skin color. For most of our evolutionary history this was certainly the case, and even today most people are raised and live among individuals who share their skin color. But by no means is this always the case. Those in the ethnic minority may find that their skin color differs from the average skin color around them, thus making their baseline skin color different from their own. If this is the case, then they may perceive their own skin to be colored. An individual of African descent, for example, who lives in the U.S., may perceive her own skin to be colored because the baseline skin color in the U.S. leans toward that of Caucasians. Similarly, if someone with a Southern accent moves to New York City, he may begin to notice his own accent because the baseline accent of his community has changed.

Our perception of various races' differences in skin color is deceptive, and this deceptive perception is potentially one factor underlying racism. In fact, there are at least three distinct (but related) illusions of racial skin color. To understand these three illusions, it is helpful to consider these illusions in the context of perceived temperature.

First, as noted earlier, we perceive 98.6 degrees to be neither warm nor cold, yet we perceive 100 degrees as hot. That is, we perceive one temperature to have no perceptual quality of warmth/cold, whereas we perceive the other to categorically possess a noticeable temperature (namely hot). This is an illusion because the physics of temperature do not account for such a large perceived qualitative difference between the two temperatures. For skin there is an analogous illusion, namely the perception we have that one's own skin is not colorful but that the skin of other races is very colorful. This is an illusion because your skin has no more or less color than anyone else's; there is no objective sense in which your skin is uncolorful but that of others is colorful. (Similarly, there is no objective truth underlying the perception that one's own voice is not accented but that foreign voices are.)

A second illusion is illustrated by the fact that we perceive 98.6 degrees as very different from 100 degrees, even though they are objectively very close. This is similar to the first illusion, but differs in that the first concerns the absence versus the presence of a perceived categorical quality, whereas this illusion concerns the amount of perceived difference in the two cases. You perceive that your own skin is very different from that of some other races, but just as 98.6 and 100 are very similar temperatures, so are the spectra underlying skin colors of different races.

Third, we perceive 102 degrees and 104 degrees as very similar in temperature, despite the fact that their objective difference is greater than the one between 98.6 degrees and 100 degrees. In skin color, we tend to lump together the skin colors of other races as similar to one another, even though in some cases their colors may differ as much from each other as your own color does from either of them. For example, while people of African descent distinguish between many varieties of African skin, Caucasians tend to lump them all together as "black" skin. (And for the perception of voice, many Americans confuse Australian accents with English ones, when the two accents are probably just as objectively different as American and English.)

As a whole, these illusions lead to the false impression that other races are qualitatively very different from ourselves and that other races are homogeneous compared to our own. It is, then, no wonder that we humans have a tendency to stereotype other races: we suffer from perceptual illusions that encourage it. But by recognizing that we suffer from these illusions, we can more ably counter them.

Waldorf Salad

My idea that color vision evolved for observing skin is a relatively new one, having occurred to me after reflecting on skin's lack of color in 2005 while a Sloan-Swartz Fellow in Theoretical Neurobiology at Caltech. But my hypothesis for the purpose of color perception is hardly the first one on the scene. For 100 years the mainstream view of why we primates evolved color vision was, as I mentioned before, that it was for finding fruit. And more recently, it's been suggested that color vision may instead have been selected to allow us to find young, edible leaves. Color is for fruits and/or leaves. For salad. Waldorf salad. These hypotheses are not necessarily at odds, however, because there could have been multiple benefits to the evolution and maintenance of our color vision. We can have our skin perception and eat with it too.

But there is reason to believe that skin perception is the dominant pressure shaping the selection of color vision. Primates vary widely in what they eat. Some eat mostly fruits, some mostly leaves, and some catch more than their share of prey. And among those that eat fruits, the kinds and colors of the fruits differ across species, as do the color changes those fruits undergo as they ripen. If color vision were selected and maintained for the purpose of sensing certain foods, then one would expect primates to exhibit a wide variety of color vision types. But among those primates with routine color vision—i.e., both males and females have it—there is very little variability in color vision despite species eating radically different meals. Radically different meals, same color vision.

With my skin hypothesis, on the other hand, there is no mystery as to why all primates have such similar color vision. Although skin colors vary across primates, we all have the same kind of blood. As we will see in the following section, no matter the primate, as blood changes in oxygenation and concentration, skin is spectrally modulated in the same ways. *That's* why we primates have the same kind of color vision.

Skin Television

Your own skin (or more exactly, the skin of your community) looks uncolored, and this is so that you are maximally able to notice changes in skin color. But what color changes can skin actually acquire?

We are familiar with many of these color changes, even if we are not always consciously aware of them. We blush with embarrassment, redden with anger, and blanch or yellow with fear. If you are strangled or choking your face becomes purple. If you watch combat sports like boxing or martial arts, you will not have failed to notice that the loser—and even the winner—often displays blue bruises, not to mention red blood. If you exercise, your face reddens; if you are feeling faint, your face may yellow or whiten. If you watch a baby clenching to fill his diaper, you will notice that his face can, in an instant, acquire a reddish, purplish tint. Babies' faces change color when they cry as well. Our sexual organs display distinct color (and size) changes when we are aroused. The skin over veins appears bluish green.

These skin colors, and the associated emotions, are reflected in culture. Angry faces are colored red, and the color is often used to indicate aggression, danger, and strength. The devil is red. In cartoons, blushing faces are pink. When women wear red, such as the stereotypical red dress, they are considered aggressive and sexy, and psychologists Andrew Elliot and Daniela Niesta found that men (but not women) do judge women who wear red as more desirable. The very meaning of red in the *Oxford English Dictionary* mentions blood and skin: "the cheeks (or complexion) and lips (as a natural healthy colour)," and "of the face, or of persons in respect of it: Temporarily suffused with blood, esp. as the result of some sudden feeling or emotion; flushed or blushing with (anger, shame, etc.); esp. in phr. red face, a sign of embarrassment or shame." Blue is often used to indicate sadness and its meanings in the *Oxford English Dictionary* include, "livid, leaden-coloured, as the skin becomes after a blow, from severe cold, from alarm, etc." Purple typically depicts extreme anger or is used for a cartoon character who chokes on food. One of its definitions also imparts sadness: "of this colour as being the hue of mourning." Green implies sickness, and the definition of green even specifies, "the complexion (often green and wan, green and pale): Having a pale, sickly, or bilious hue, indicative of fear, jealousy, ill-humour, or sickness." Cowardly faces are often colored yellow, and yellow's meanings include "craven, cowardly"; but yellow is also associated with happiness. Figure 4 shows four emoticons used in Web postings that help writers better communicate their mood, and Figure 5 summarizes meanings of several color terms that refer to skin, blood, and emotion.

Clearly, skin can achieve a wide variety of colors. How does our skin do this? And what exactly are its limits? As it turns out, skin has the somewhat magical ability to dynamically acquire any hue at all!

The key to this special ability is blood. Specifically, two features of blood matter here: (i) the quantity of blood in the skin, and (ii) the oxygenation level of that blood. As we will see later, our color vision is able to sense these blood qualities—but how do changes in these two features change the color of skin in the first place? If there is less blood under the skin than normal, skin appears yellowish. If there is more blood than usual, skin appears bluish. If the blood is more oxygenated than normal, skin appears reddish. If it is less oxygenated than usual, skin appears greenish. Figure 6 shows the palm of an undergraduate student named Qiong "Gus" Zhang who helped with the generation of this figure while I was at Caltech.

In other words, whether skin appears yellow or blue depends on how much blood is under the skin. Skin becomes yellowish if you squeeze the blood out of it. Try pushing the blood out of your palm, and you will notice that the pressed spots appear yellow until the blood rushes back in. Or simply make a tight fist and look at your knuckles. You should see yellowish spots where the skin presses against the bone. Without blood, the skin not only appears more yellow, but also lighter (and thus whiter). Skin becomes bluer (and also darker) when the amount of blood in the skin increases. One easy place to find skin with an increased concentration of blood is in your veins. The skin over visible veins appears bluish (but also has a greenish tint because the blood in veins is low in oxygen). Another way of seeing bluish skin is to cut off the circulation to your hand by, for example, tightly grasping your wrist with your other hand, tourniquet-style. After a minute or so, blood will accumulate, and your hand will begin to appear bluish. (Your hand will also take on a reddish tint because, unlike the blood in your veins, the blood that accumulates here has a higher oxygen content. The mix of the two colors is why the skin appears purple.) Since bluer skin is a sign of excess blood in the skin, which can be due to poor circulation, perhaps this is why blue is so often associated with lethargy and sadness.

Whether skin is red or green depends on how much oxygen is in the blood underneath it. There is no simple way to raise and lower the oxygenation level of the blood beneath your skin, but one way to observe the red-green changes is by comparing your skin's color in the two cases we already discussed: over veins and when using a tourniquet. In each case, there is more blood than normal, but the color of the former is greenish-blue, and the latter is reddish-blue. The difference is that the former involves less oxygenated blood, and the latter involves more oxygenated blood. Anemics tend to have a greenish hue to their skin because their blood is low in oxygen. This may help explain why green is sometimes associated with weakness (because lack of oxygen makes one weak), whereas red may be associated with strength (because plentiful amounts of oxygen in the blood make one stronger).

We have seen that skin can take on many colors, and we have discussed how blood is what makes this possible. See Figure 7 for a summary. Modulating the amount of blood in the skin shifts skin's color between yellow and blue, and modulating the blood's oxygenation level shifts skin's color between red and green. From these seemingly tepid observations something extraordinary follows: Skin can acquire any hue at all! Why? Because every hue is created from a combination of primary colors—blue, green, yellow, and red. Skin can become yellow-green by lowering the amount of blood and decreasing the level of oxygenation. It can become blue-green by raising the amount of blood and decreasing the level of oxygenation. It can become blue-red—or purple—by raising the amount of blood and increasing the level of oxygenation. And it can become yellow-red—or orange—by lowering the amount of blood and increasing the level of oxygenation. These two changes in blood—in the quantity and oxygenation—are sufficiently powerful to allow skin to appear to be any hue at all.

It is important to realize that when we see colors on a section of skin, it doesn't mean we would necessarily perceive that piece of skin to be the same color if we saw it all by itself. This is because, similar to perceived fever and perceived accents, an object's perceived color depends crucially on how that object's spectrum *deviates* from the surrounding baseline spectrum. You perceive your veins to be blue-green, but they only appear to be blue-green next to your normal baseline skin. If you were to look at your vein through an aperture or slit such that you could see *only* your vein, and not any of the baseline skin nearby, your vein would no longer appear blue-green. It would instead appear approximately skin colored (e.g., peachish, tannish, etc.). Your vein differs from your baseline skin color in that it is both a little greener and a little bluer, and that's all your brain needs in order to see it as blue-green. This is important for understanding how our skin changes color specifically for the purpose of signaling to others. When we blush, for example, some regions of our cheeks take on a red tint. In order for others to see this as red, however, it is crucial that some of the surrounding skin remain at baseline. If the entire face were to become identically red, then unless you actually watched the dynamic change occur, you would not see the resultant cheeks as red because they wouldn't be redder relative to the surrounding skin.

What is the significance of skin's color-changing ability? It is an indication that color vision is designed for seeing skin color changes. To understand why, ask yourself: How many other natural objects can dynamically achieve every hue and *also* appear to have no color? Fruit can go through several hues over the course of a week as it ripens, and leaves can also display multiple hues while maturing. But neither fruit nor leaves can hit every hue, much less do so in such a dynamic manner, able to pass back and forth *between* hues. The only other objects in the natural world able to display multiple hues are the skins of just a few other animals, such as the chameleon and the squid. One group of researchers led by biologist Ruth Byrne has classified the colors dynamically obtainable in a reef squid as "pale, white, yellow, gold, brown, and black." Other cephalopods are able to appear blue, but only in metallic shades. The skins of these animals do not, however, appear able to display a continuum of hues, nor can they return to baseline or "turn off" like human skin, at least not to human eyes (e.g., squids at baseline appear gray, and chameleons at baseline appear green).

An object capable of dynamically displaying all possible hues and also appearing uncolored is, then, a rarity in the natural world. And if you *do* find an object that is capable of reaching all the hues and also returning to a baseline, then it is very likely to be by design, whether human or evolutionary. It suggests that it is no accident that skin can display all the hues and has an "off" switch. Skin is a full-color monitor by design—design by natural selection.

When women put on make-up, they're explicitly engaging in color signaling. Some of the colors they apply are intended to cover up what they perceive as blemishes and are skin-colored—aptly called "base." Some of the colors they choose are purposely colorful, such as blush for the cheeks. Paradoxically, such conscious efforts at color signaling actually serve to mask the natural color signaling the face evolved to do and the eyes evolved to sense. Putting on make-up is like placing a colorful photograph in front of your television set—it provides a colorful view, but it entirely blocks the dynamic and meaningful colorations occurring on the screen. I've never been a fan of make-up on women, vaguely feeling that much of what's important in a face is missing when it's covered with make-up. The research on color perception and skin supports this intuition of mine: too much make-up takes full-color skin television and simply turns it off.

The analogy of skin to a color monitor suggests that skin is capable of achieving all the hues because of selective pressure. But there is a problem with this. These skin color modulations via blood are not new to primates but rather have been around throughout the history of mammals. Therefore, it is *not* the case that skin has been modified by natural selection to achieve all the hues (other than, perhaps, evolving capillaries nearer to the surface of the skin in some regions in order to better *display* the color modulations). We are wearing the same old skin and have the same ancient blood as our mammalian ancestors. So if our blood and skin have no new powers of color, then how is it that skin is able to achieve all these colors?

The answer is that our color vision evolved in response to our skin's natural properties, not vice versa. Our skin didn't change to suit our eyes, but rather, our eyes changed in order to better see our skin. Skin's spectral properties are modified the same way they were tens of millions of years before color vision ever came about. What happened when primates evolved color vision was that they acquired perceptual sensing equipment in the eye and brain that were specifically designed to pick up on these ancient spectral changes. Skin was not originally a full-color monitor, at least not until primates evolved to possess vision that let them see skin in full color. It is actually in the eye and brain—and the color vision they give us—where we find the magical source of our skin's powers of color, not in the skin at all!

Persons of Color

When I tell people about my hypothesis for the evolutionary function of color vision, one out of every two people asks me, "What about those with very dark skin? Do they show these color modulations?" Sometimes the question is in the spirit of academic criticism: at some point all humans had very dark skin, so if skin color signals don't show on dark shades of skin, then color vision must not be for seeing skin coloration. However, even if it were true that color signals cannot be seen on very dark skin, color vision evolved long before humans did, and our ancestral primates may well have had lighter skin. (Primates with color vision have a wide variety of skin colors, as you can see for yourself by looking ahead to Figure 9.) If this were the case, our color vision would not have been particularly beneficial while we had dark skin, but it would have become useful as some humans spread around the globe and gradually acquired lighter skin.

Many of those who ask the question aren't interested in academic points but are merely curious. Darwin was also curious about this. In his book *The Expression of the Emotions in Man and Animals,* he writes,

> Several trustworthy observers have assured me that they have seen on the faces of negroes an appearance resembling a blush, under circumstances which would have excited one in us [Caucasians], though their skins were of an ebony-black tint. Some describe it as blushing brown, but most say that the blackness becomes more intense (318).

He follows this with,

> It is asserted by four of my informants that the Australians, who are almost as black as negroes, never blush. A fifth answers doubtfully, remarking that only a very strong blush could be seen, on account of the dirty state of their skins. Three observers state that they do blush; Mr. S. Wilson adding that this is noticeable only under a strong emotion, and when the skin is not too dark from long exposure and want of cleanliness. Mr. Lang answers, "I have noticed that shame almost always excites a blush, which frequently extends as low as the neck" (319–320).

Concerning such observations, one must take into account that many of these observers were Caucasian and likely raised in an environment where lighter skin was the baseline. We would expect such individuals to be less capable of discriminating skin color deviations around a dark-skinned baseline. For example, Darwin writes that,

> Von Spix and Martius, in speaking of the aborigines of Brazil, assert that they cannot properly be said to blush; "it was only after long intercourse with the whites, and after receiving some education, that we perceived in the Indians a change of colour expressive of the emotions of their minds" (318).

Although the implication is that the aborigines' education led to greater blushing, it is much more plausible—to put it mildly—that it was the whites' experience with aborigine skin color that eventually trained them to perceive it.

Skin color variations *are* visible on the darkest skin, and even Caucasians with less exposure to darker ranges of skin colors can notice the changes. This is because, whether dark or light, human skin possesses a reflectance spectrum with the same signature features (see Figure 3), and later we'll see that the spectrum changes in the same way as the two blood variables, concentration and oxygenation, vary.

Naked. Who? Where?

We see bare faces every day of our lives, and we have evolved for millions of years with these faces, so they are as natural to us as having two eyes. When we see other primates with bare faces, we do not raise an eyebrow. We *do* bat an eye when we see a completely furry human face, however, and are willing to pay money at a carnival to gawk at one (see Figure 8). This familiarity with bare faces and bare skin puts us in a very poor position to appreciate just how unusual it is to have a bare face.

The typical mammal has a furry face. The ancestral primate was just as hairy, and only some of today's primate species have come to have significant bare spots on their faces. We bare-faced primates are the oddballs, and our furry ancestors would be ashamed at all the skin we show. Adding to the embarrassment is the fact that we have a tendency to have other naked spots as well, such as on the rump and genitals, and sometimes the chest. Humans are the worst violators of all, having nearly rid ourselves of hair altogether.

What explains this mystifying exhibitionism? Why do most mammals have furry faces while we and a selection of our primate brethren have bare spots? In light of the skin color hypothesis we have been discussing, an idea naturally presents itself. You can see color modulations on bare skin, but not on furry skin. Therefore, perhaps it is those primates with color vision that also have bare spots, implying that the bare spots are for color signaling. That is, there is no sense in evolving color vision that turns skin into a full-color skinevision monitor if there is no bare skin to watch movies on.

And, indeed, the primates with color vision are the ones with the bare spots on their faces, while the primates without color vision have typical mammalian furry faces. Figure 9a shows representative primates without our full color vision; they're clearly furry. Within the group of primates who have color vision, we see two different types. In New World Monkeys, shown in Figure 9b, only the females have color vision. In Old World primates like us, shown in Figure 9c, both males and females can see in color. Both of these groups have bare spots, as can be seen in the figures. And, among the Prosimian primates, which usually do not have color vision and are furry-faced, the two examples that *do* have color vision break the Prosimian mold by having bare faces (the top two photos in Figure 9b).

I should emphasize before moving on that we don't expect primates with color vision to have colorful skin—something researchers and journalists sometimes suggest. In fact, they often propose that if color vision is designed to see skin, then primates with color vision should have permanently colorful skin; usually they suggest red. Some primate species with color vision do appear red to us, like the red-faced Macaque and Uakari, but that just means their baseline skin color is different from ours. (There are *numerous* different baseline skin colors, even within our *own* species.) As we discussed earlier, *they* probably don't perceive themselves as red, but rather, perceive themselves as lacking color, thereby helping them discriminate color variations from their baseline. If they did see themselves as red, then they would be much less able to notice the small color modulations color vision is designed to detect. Color vision is no more about red skin than perception of accent is about Southern drawls. Skin that is *bare*—not skin of any particular color—is what we expect to see in primates with color vision, and that's what we find.

Bare skin, then, is for color signaling. Actually, one must be a bit more specific. After all, early furry-faced primates had bare palms; even your pet dog has bare skin on his paws. Really, all the bare spots where the typical furry mammal is furry—the "new" bare spots—appear to be meant for color signaling. And indeed, all of the new bare spots are on parts of the body that are easily seen by others, such as the face and rump.

But if *this* is true, then what does it say about our overly bare bodies? It suggests that our (new) bare spots are for color signaling, answering Desmond Morris's question about why we are the (most) naked ape. It only makes sense to have bare skin on parts of the body that can be seen by others, and once we began walking upright, our fronts and backs could be seen much better than before. Perhaps, then, fur was reduced to better enable color signaling, and we are the lone naked ape because we are also the lone bipedal color-vision primate.

We are not *entirely* naked, and the locations of the spots where we universally retain fur fit nicely with the idea that our naked spots are for color. If a part of the body cannot be seen by others, then there is no reason to color-signal there and so no need for that part of the body to be bare. In short, if a body part cannot be seen, then we expect it to be furry (ignoring the ancestrally ancient bare spots such as the palms and the bottoms of the feet). Is this true for us? There are three parts of our body that cannot be seen, and thus would not be good candidates for color signaling: the top of the head, the underarms, and the groin. And notice that these are the universally furry spots on humans. The fact that we still have fur in places incapable of color-signaling suggests that the (new) bare spots are for the eyes of others. In fact, when the groin is a counterexample to this—when the genitalia engorge and bare skin becomes visible there, as in humans—it is in circumstances where another person *is* likely to see this skin.

Although it's true that "if a body part cannot be seen, then it is furry," we do *not* necessarily expect that "if a body part *can* be seen, then it is bare." There are a number of potential advantages to fur, and these benefits can counter the loss of some of our color-signaling canvas. For humans, male facial hair is the best example, probably useful for sexual selection (eyebrows, probably useful for exaggerating facial expressions, are another). But even here, notice how male facial hair does *not* look like the facial hair of the fellow we saw earlier with hypertrichosis in Figure 8. Male facial hair is confined to the beard and mustache regions (that is, male facial hair is kept away from the color signaling spots), so that even someone as hairy as Santa Claus can color signal effectively—thus the songs about Santa's rosy cheeks and cherry nose.

One thing left unanswered in this discussion is why we would have evolved to signal our emotions with color in the first place. What do skin-color signals add that couldn't have been communicated by facial expressions or gestures? One answer is that colors are displayed without the need for any muscular activity, freeing up the muscles to carry out whatever behavior the animal wishes. Try maintaining your angry face while eating. Or, if you're a female chimpanzee, try holding a come-hither expression for the entire time you're in heat. Also, color signals are displayed on parts of the body where there are no muscles capable of gestures, like the chest and rump. This not only gives us a greater amount of canvas on which to signal, but allows signals to be located right in the region that matters. Female come-hither looks may be effective in attracting male attention, but an engorged red rump leads the males to the appropriate place.

Color signals can also more directly communicate the physiological state of an animal to others than muscle expressions. "*Not only am I angry at you, but do you see all this red in my face? It means that I am in good physiological condition, well-oxygenated, not winded, and ready to fight.*" You can't fake oxygenation, at least not for long. The same may be true for other color signals as well. Even a simple blush requires good physiological condition, not to mention healthy skin.

Another important advantage to color signals over muscle-driven expressions is that color signals are inherently harder to manipulate. By their nature, most muscles must be manipulated consciously by the animal. Blood oxygenation, however, has more serious responsibilities, like keeping every cell in your body alive. Having an animal's fundamental cardiovascular characteristics be separate from his volition—no animals allowed in the control room—is therefore a wise evolutionary choice. One consequence is that colors can be signaled even when an animal is asleep or unconscious, particularly in the case of breathing difficulties, especially in infants—a topic we'll talk more about in the following section. More generally, color signals tell us how the animal *really* feels, something that was probably fundamental to the evolution of reciprocal altruism, a concept developed by evolutionary biologists George Williams and Robert Trivers. Living in a community of altruists can be beneficial for each of the individuals, but only so long as cheaters are not allowed to live and prey on their fellows. If cheaters can prosper, the community will soon be dominated by cheaters. Robert Trivers has argued that many of our emotions serve to maintain the conditions needed for reciprocal altruism. Anger signals to the cheater that you've caught him and that he should expect punishment. Grudges enforce punishment upon a cheater for a longer period of time. And if discovered cheating, signals like blushing are a means by which one can signal that one is truly contrite. As psychologist and linguist Stephen Pinker has eloquently speculated, such emotional signals wouldn't serve to enforce reciprocal altruism if the emotions could be turned on and off like light bulbs. If the victim could *decide* whether or not to be angry, then it may be in his best interest not to punish the cheater because the cheater may not *always* cheat, and could be of some use in the future. The cheater, however, would know this and thus not be deterred from cheating. But if cheaters expect anger to be the automatic, unstoppable consequence of being cheated, cheaters will be deterred. And if someone accused of cheating could simply switch on contrite emotions at will, then his accusers would have little reason to believe he is sincere. However, if people can expect blushing as an automatic, unstoppable consequence of true regret (the response being difficult to consciously fake), then the accusers are more likely to accept the implicit apology.

A Healthy Glow

As I've been saying, our power of color telepathy is useful for reading the emotional states and moods of others. But it is also useful for recognizing when others are sick or in acute distress. Many diseases, disorders, and injuries lead to changes in the oxygenation or concentration of blood flow to our extremities. These blood-related changes lead to changes in skin color that our eyes are able to detect. Generally, poor circulation leads to excess concentration of blood, which look bluer (and darker). Skin that is experiencing loss of blood looks yellowish (and lighter). Deoxygenated skin becomes greener, which is crucial for understanding central cyanosis, a common clinical condition indicating acute deoxygenation. Deoxygenated arterial blood leads to a reaction in the capillaries that results in greater pooling of capillary blood, so that skin that typically looks purplish (as in the tourniquet of Figure 6), now shifts greenward to blue. We also bruise when injured, which leads to a characteristic sequence of colors as the bruise heals, but these color changes are ruled by a mechanism other than blood oxygenation and concentration. Generally, anywhere a pulse oximeter is used within a hospital is a place where our eyes—and the oximeters we carry within them—can be useful (if not as sensitive) to keep tabs on a patient's condition.

An undergraduate student at the Rensselaer Polytechnic Institute named Kevin Rio spent a semester with me studying the extent to which the field of medicine explicitly mentions skin color as a diagnostic sign and found that approximately 10 percent of all disorders do so. The clinical disciplines in which skin color plays a significant role are emergency medicine, pediatrics, cardiology, and OBGYN. (It goes without saying that skin color is critical in dermatology, although the source of color here is typically rashes and other skin changes, not underlying acute blood changes.) Skin color tends not to matter as much for otolaryngology, psychiatry, orthopedics, and internal medicine. One crude measure of the importance of color for a field of medicine is the percentage of books in the discipline that come up using "color atlas" as a key word in Google Book Search. In emergency medicine, pediatrics, cardiology, and OBGYN, the percentages are 7.3 percent, 3.8 percent, 4.6 percent, and 4.8 percent, respectively, whereas the term comes up, respectively, only 0.3 percent, 0.5 percent, 1.3 percent, and 1.4 percent of the time in otolaryngology, psychiatry, orthopedics, and internal medicine. The disciplines in the first group are also more likely to use pulse oximeters than the latter group—of the four clinical disciplines in which skin color matters, 18.1 percent, 6.1 percent, 12.1 percent, and 3.1 percent of the books in the respective fields mention oximetry, whereas in the four disciplines where skin color is not as critical, the respective percentages are 1.2 percent, 1.2 percent, 2.4 percent, and 2.5 percent. This suggests that clinical medicine has evolved over the years to harness the capabilities of our natural oximeters, even without realizing that our eyes had actually evolved for that purpose.

Physicians with normal color vision, then, are using their eyes as oximeters. If this is the case, then we should expect color-blind physicians to lack this clinical power. In fact, medical doctors have noticed that color-blind physicians are severely handicapped when diagnosing patients. Dr. Heinz Ahlenstiel writes that

> Slight reddening of the skin, as in blushing, is overlooked by the red-green blind. Growing pale is also overlooked, as is a very slight scarlet rash. Stronger reddening of the skin is labelled as dark grey shadow by the red-green blind. It is in this manner that the inflammatory lymph streaks associated with blood poisoning are recognized. Reddening of the interior parts of the body, in the throat, nose, ears and epiglottis, are more difficult to recognise. The bluish discolouration of the lips and nails in circulatory disorders remains imperceptible. Blood spots are imperceptible to the red-green blind on dark materials.

Dr. Anthony Spalding describes how one color-blind doctor

> had failed to see the extreme pallor of a woman waiting for surgery. "Anyone could see it," the gynaecologist said but I could not. The operation was delayed for a week while the patient received a blood transfusion.

Another color-blind doctor Spalding interviewed responded as follows:

> I did a year in pathology but did not realise that stains showed different tissues—no one told me. My eyes pick up very fine physical features of rashes et cetera. I feel this and body language are major advantages I have developed. I frequently ask my colleagues for advice, especially over babies with rashes and fevers et cetera, and the chance of a red ear or throat. I feel very vulnerable at the end of a busy surgery. I believe there are times when patients describe red rashes and the nurse points to the invisible spots. I cannot believe that I could be guided to two careers, which depended so heavily on colour, namely, pathology and dermatology. The profession offers no guidance and I feel vulnerable at times.

And the optometrist Dr. David M. Cockburn begins one of his papers by reflecting, "As a child I could not understand what people meant when they said someone was blushing," also complaining about his "embarrassment when a patient complained of a red eye but the offending side . . . was not specified," and that his "most severe problem was in differentiating between blood and pigment in the retina."

The color blind, then, not only lack the ability to sense the emotions and moods of others, but also lack the power to sense those who are sick. I have wondered on occasion whether this could be one reason color blindness is more common among men than women. Almost 10 percent of men are color blind, but less than half a percent of women have this problem. In fact, *only* female New World Monkeys have color vision—all males are color blind. In humans, not only are females rarely color blind, but some may even possess *extra* powers of color, having four cone types rather than the three for normal color vision. I can imagine telling lots of stories about why color vision may have been under stronger selection pressure for females, but one speculative hypothesis worth mentioning concerns the clinical benefit it provides, especially for infants. With two young children in my life, I have been struck by the extent to which every cough, sneeze, clench, choke, or cry changes the color of their little faces. Their color signaling seems to be amplified. This suggests that infants whose mothers couldn't detect these signs would be at greatly increased risk. An ancestral mother who could notice the signs of choking could administer the Pleistocene epoch's Heimlich maneuver (rapid upside-down shaking); the color-blind female would not even have noticed her infant was in trouble.

Color blindness is not the only way to have a handicapped clinical sense. As we discussed earlier in this chapter, we perceive our skin as uncolored, and this is crucial for helping us perceive skin color changes around a baseline. We also discussed how our ability to perceive skin-color modulations is greatly reduced when it comes to those of different skin colors. In addition to having implications for the psychological foundations of racism, this suggests that clinicians will be less capable of noticing clinically significant color signals in individuals of a race different than their own. For example, an emergency room doctor of African descent who moves to Provo, Utah, is likely to be somewhat blind to the clinical color signals of the locals until his brain adjusts to the local baseline color. Therefore, one can be effectively color blind to clinical skin color changes without being color blind.

Thus far in our discussion of our clinical color sense, I have been implicitly assuming that when we're sick or injured, there are certain inevitable skin color changes that occur and are noticeable on bare skin. In other words, I've been implying that your furry dog undergoes the same skin color changes humans do. However, it's possible that we have been selected to color signal in order to show others our troubled state. We know this is true for emotions and moods: blushing is a programmed color signal, not an accidental side effect of underlying physiological processes. There is no reason to think this is not the case for some clinical color signs as well. Choking infants who displayed a very clear choking color would have had a greater tendency to be saved by Mom. Over time, infants would have evolved better and better mechanisms to bring Mom running to their aid. Because puppies are furry, they don't receive any benefit from color signaling (even if a puppy's mother could see in color). If you shave a puppy, you're likely to see much less color modulation than you do on human infants. This could even be the case for bruises. Once an animal is bare-skinned and begins to color signal, natural selection could cause specific injured spots to become more visible: bruises could be more visible on our bare skin than on a dog's bare skin. Veterinarians assure me that bruises on dogs, horses, and cows are visible when the animals are shaved, but we'd need a properly controlled test that compared similarly thick skin to similarly traumatic impacts to judge the level of color change accurately. Most animals don't end up in the animal hospital for a bruise unless the bruise is a very serious one. My dogs used to tumble through the rocks in creeks all day chasing balls, and if I had suffered even a hundredth of the impacts they did, I'd have been black and blue for days. I've never shaved them to check for bruises, though next time I just may do so.

Our color vision therefore not only gives us the power to read emotions and moods, but also gives us the ability to sense illness and injury. Both of these probably contributed to the evolution of color vision and the co-evolutionary loss of fur.

What It's Like to See in Color

Thus far we've talked a lot about skin but very little about the eye, other than concluding earlier that the source of our empathic—and clinical—power lies in the eye and brain, not in the skin itself. The eye is able to turn run-of-the-mill skin and blood into a full-color monitor. But before we can make sense of this, we need to understand how we perceive color and how those perceptions are organized.

In attempting to understand our color perceptions, the rainbow might seem like the place to start. After all, don't we see all the colors when looking at a rainbow? One can easily see many of the colors in the rainbow, namely the "ROYGBV" colors: red, orange, yellow, green, blue, and violet (Figure 10, top). Furthermore, the rainbow displays all the shades in between, such as reddish orange and yellowish green shades. That's a lot of colors. But is it all the colors? As schoolchildren we were taught that white light contains all colors, and that putting a prism (or drops of rain as in a rainbow) in front of light "pulls apart" the white light, revealing light of all possible colors arrayed in a rainbow-like spectrum. This would indeed suggest that rainbows possess all the colors. In fact, the word "spectrum" in the English language even connotes the idea of possessing all possible variety. The definition in the *Oxford English Dictionary* is, "the entire range or extent of something, arranged by degree, quality, etc."

However, a moment's observation reveals that the rainbow does not possess all the colors. Look around the room you're in. You should have little difficulty naming most of the colors around you, and yet most of the colors you see probably do not actually look like the ones in the rainbow. Sure, the blue of the sky, the orange of an orange, and the red of fresh blood might look approximately like colors found in a rainbow, but what about purple, brown, pink, gray, khaki, and maroon—not to mention skin color? These latter colors are not in the rainbow. The rainbow captures only one dimension of our color perception; it misses so many colors because our color perception has *three* dimensions, not just one.

What the rainbow does capture is hue, although it doesn't even capture hues perfectly because it misses an entire range of them, namely the purple ones. The purple hues are simply not in the rainbow, meaning that no single wavelength of light appears purplish. In order for incoming light to appear purple, its spectrum must have peaks at both the low wavelength and the high wavelength ends. The purple hues are perceptually in between blue (or violet) and red, but the rainbow has blue and red at opposite ends—they do not combine anywhere within the rainbow. (Violet is not the same color as purple; the former refers only to the extreme color in the rainbow beyond blue.) We can "fix" the rainbow by attaching purple hues to the outside edge of the red hues, but this would still downplay a crucial feature of hue, namely its "shape." Hues should not be depicted along a straight line at all (as shown in the upper part of Figure 10). Instead, hues should be shown in a circle. The bottom part of Figure 10 illustrates this, where the red and violet parts of the rainbow are glued together, using the purple hues to fill the gap. Purple hues begin with a purple-red, turn slowly into more central purple, and eventually become a very purple-blue, nearly violet.

Hue, however, is just one of three dimensions of color. The other two are saturation and brightness. Saturation concerns how "full" of color a particular hue is. For example, gray, before you add any red hue to it, has zero saturation of red. The more red is added, the more saturated with red the color becomes. And as you de-saturate any hue, the more gray it becomes. This last point is crucial in understanding the "shape" of saturation. Figure 11 is akin to the depiction of hue in Figure 10, but instead of just considering the circumference of the circle—the hues—Figure 11 shows a flat disk. The disk has gray at the center, from which any hue may be achieved by gradually increasing saturation in the appropriate direction. Now we have two dimensions of color rather than one: hue, or the location on the circumference around the disk, and saturation, or the distance from the center of the disk.

The third and final dimension of color is brightness. At first glance, this may not seem to be a dimension of color per se, but it is a crucial aspect of our experience of color. If you keep a color's hue and saturation fixed while changing its brightness, there will be a perceivable color change. For example, the color brown is actually a reddish hue, with low saturation and low brightness. Brown will no longer appear brown if you increase its brightness (it will appear red or orange instead). And clearly, gray qualitatively changes considerably as you vary the brightness, becoming white or black when the brightness is sufficiently raised or lowered.

Thinking about color in terms of hue, saturation, and brightness is very important, for without understanding these three color dimensions, one may not be said to properly understand color at all. Still, we are lacking many details. For example, where exactly are each of the hues located around the circular disk, and why? In Figures 10 and 11, I have shown red as opposite of green and blue as opposite of yellow. But nothing we have discussed thus far has given us any idea that this might be the case. Red could be opposite blue, with the purple hues spanning one half of the disk, and all the blues, greens, yellows, and oranges crammed into the other half. We will see that there is another way of thinking about this color space, one that is informative and helps address where exactly the hues are situated around the perimeter. It will also be crucial for making sense of the colors we primates can see but other mammals cannot. This alternate way of thinking about color space is the result of some crucial discoveries by the great nineteenth- and twentieth-century thinker Ewald Hering (and has since been verified by vision scientists Leo Hurvich and Dorothea Jameson in the 1950s, and Robert Boynton, among many others).

Hering first noticed that nearly all the hues around the hue circle appear to be mixes of two other hues. For example, purple appears to be a combination of blue and red, and orange a mix of yellow and red. From this, something profound follows: there must be some hues that do not appear to be a mix of others at all. There must be pure, or primary, hues. Suppose that hue B appears to be a combination of two other hues, C and D. Are C and D themselves mixes? If so, then B would actually appear to be a mix of more than two hues. For example, if C is a mix of E and F, and D a mix of G and H, then B would be a four-way mix of E, F, G, and H. However, Hering discovered that each hue appears to be a mix of at most two hues. Therefore, if B is a combination of hues C and D, then C and D cannot be mixtures of other hues. They must instead be a special kind of perceptually pure (primary) hue.

Which hues are perceptually pure, and how many are there? Hering noticed two things in this regard, each leading to the conclusion that there are four primary hues rather than two, or ten, or a hundred. First, he observed that there were only four hues that appear perceptually pure, or unmixed: blue, green, yellow, and red. If these four are the only pure hues, then whenever a hue appears as a mix, it should appear as a mixture of two of these four hues. Hering's second observation further supported that this was indeed the case: all the mixed hues were indeed combinations of two of these four unitary hues. Orange is a mix of red and yellow, for example. This amounts to an amazing discovery because it establishes these four primary hues as the perceptual building blocks underlying all other hues. The infinite subtle hue variations around the circle—from violet into blue, green, yellow, orange, and red, into purple and back again to violet—can be simplified to these four pure hues. I hope you are now beginning to see why hue, saturation, and brightness are not sufficient for understanding our internal experience of color. Thinking in terms of these three dimensions alone gives us no insight into the fact that there are just four perceptually pure hues, and that all other hues are built from pairs of these primaries.

Hering noticed something even more surprising, something that few people know about their own experience of color. Hues—and more generally, colors—have perceptual opposites. No one will be surprised to learn that black and white are perceptual opposites, but what is the opposite of red? At first glance, it seems to be a nonsensical question, and personal reflection does not easily help us answer. One would imagine that opposite hues would lie on the opposite sides of the disk in Figure 10, but how do we know which colors are on opposite sides? Recall that there are just four primary hues, and that every other hue is perceived as a mix of exactly two of these primaries. Hering further noticed that some combinations of the four primaries never occur. Red can mix with blue to create purple, and red can mix with yellow to make orange. But he noticed that red cannot mix with green to make a new hue. That is, he discerned that there is no hue that can be described as reddish-green. Similarly, he noticed that although blue can combine with red or green, mixing yellow and blue does not make a new hue. That is, there is no hue that can be described as yellowish-blue.

Some of you might be protesting that green is yellowish–blue, because we all know that "yellow and blue make green." Physically adding yellow paint to blue paint does tend to result in a color we see as green, but our perception of pure green cannot be described as something that "perceptually feels" like a mix of yellow and blue. Purple, on the other hand, *does* "perceptually feel" like it contains both blue and red, and orange feels like it contains both red and yellow. The point here is that there is no color that perceptually feels like it contains both yellow and blue—green feels as if there is neither yellow nor blue in it—just as there is no color that feels like it contains both red and green.

So, red does not perceptually mix with green in the head, and blue does not mix with yellow in the head. And this means that there are only four possible primary hue combinations in perception: blue-green, green-yellow, yellow-red, and red-blue. The remaining two combinations, blue-yellow and green-red, simply never occur. What are we to make of this? Why can some primary hues perceptually mix, yet some others cannot? Hering's conclusion was that blue and yellow, and red and green, must be perceptual opposites. Why would he conclude this? One fundamental property of opposites is that they can't be usefully combined. For example, a man can be tall and happy, but he can't be tall and short, nor can he be happy and sad. The fact that blue and yellow cannot usefully combine as a mix to make any other hue suggests that blue and yellow are perceptual opposites. The same is true for red and green. Therefore, blue and yellow must be on opposite sides of gray in the color disk in Figure 11, and so must green and red. This is an important start: we know where blue is relative to yellow and where red is relative to green. But where within the perceptual space of the color disk should red and green be placed in comparison to yellow and blue? Let's consider first where red should go, relative to yellow and blue. Red is a pure hue, containing no yellow or blue, and for this reason it is equally dissimilar to both yellow and blue (i.e., it has no similarity to either). Red should therefore be placed equally distant from yellow and blue on the perimeter of the hue wheel. And because green lies opposite red, green also lies equally distant from yellow and blue, but on the opposing side. What we now have are blue, green, yellow, and red placed at 90-degree intervals around the hue wheel (the illustrations have thus far been showing it like this, in anticipation of this conclusion). All the mixed hues thus lie in one of the four quadrants of the disk, as shown in Figure 12.

Now that we know where the colors are situated on the hue wheel, it is possible to introduce a new way of thinking about our internal perception of color. This will be crucial to understanding why it is that we primates see a superior range of colors compared to other mammals. Rather than think about the color disk in terms of the location around the circumference (hue) and the distance from the center of the disk (saturation) as in Figure 11, we can think of it in terms of two, simpler straight-line axes, akin to x- and y-axes, as shown in Figure 12. One axis is the blue-yellow axis, a line connecting pure blue to (its opposite) pure yellow, passing through gray in the center. This is shown in Figure 12 as the vertical axis, or the y-axis. One may consider gray as being at zero along this axis, with blues as positive numbers and yellows as negative numbers (yellow is "negative blue"). The second axis is the red-green axis, a line connecting pure green to (its opposite) pure red, passing through gray in the center. This is shown in Figure 12 as the horizontal axis, or the x-axis. Along this axis, gray is again at zero, with reds as positive numbers, and greens as negative numbers (green is "negative red"). That is, rather than via hue and saturation, it is possible to describe these two color-disk dimensions through two fundamental

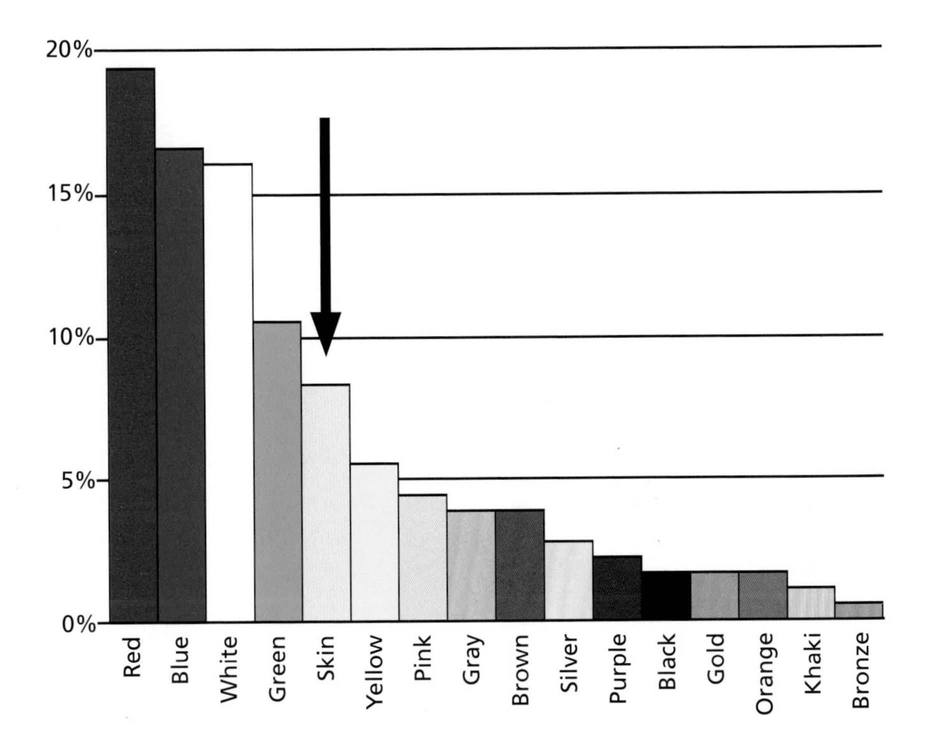

FIGURE 1. *Proportions of colors I measured in 1,813 pieces of clothing from* Racinet's Full-Color Pictorial History of Western Costume: With 92 Plates Showing Over 950 Authentic Costumes from the Middle Ages to 1800. *The arrow indicates the data point for "skin-toned" clothes.*

FIGURE 2. *List the colors you see in this picture.*

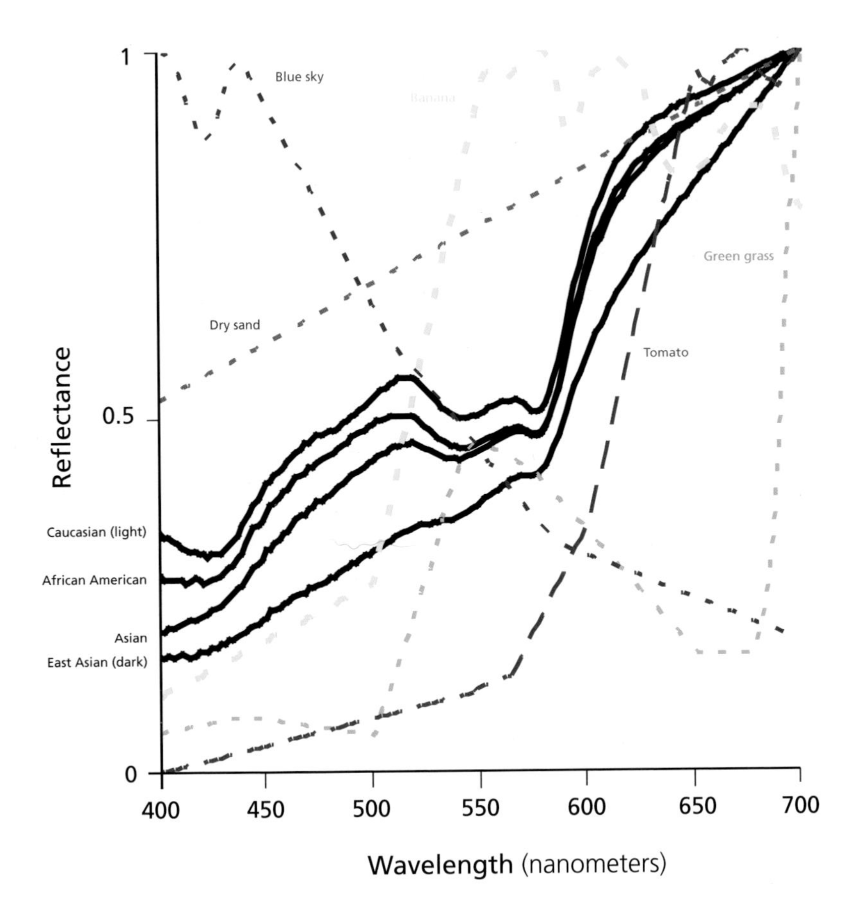

FIGURE 3. *Reflectance spectra from a variety of human skin (data from NCSU spectral database). Notice how similar they are, compared to a variety of other spectra.*

FIGURE 4. *Example "emoticons" with colors. Green is often associated with sickness, blue with sadness, red with strength or anger, and yellow with happiness.*

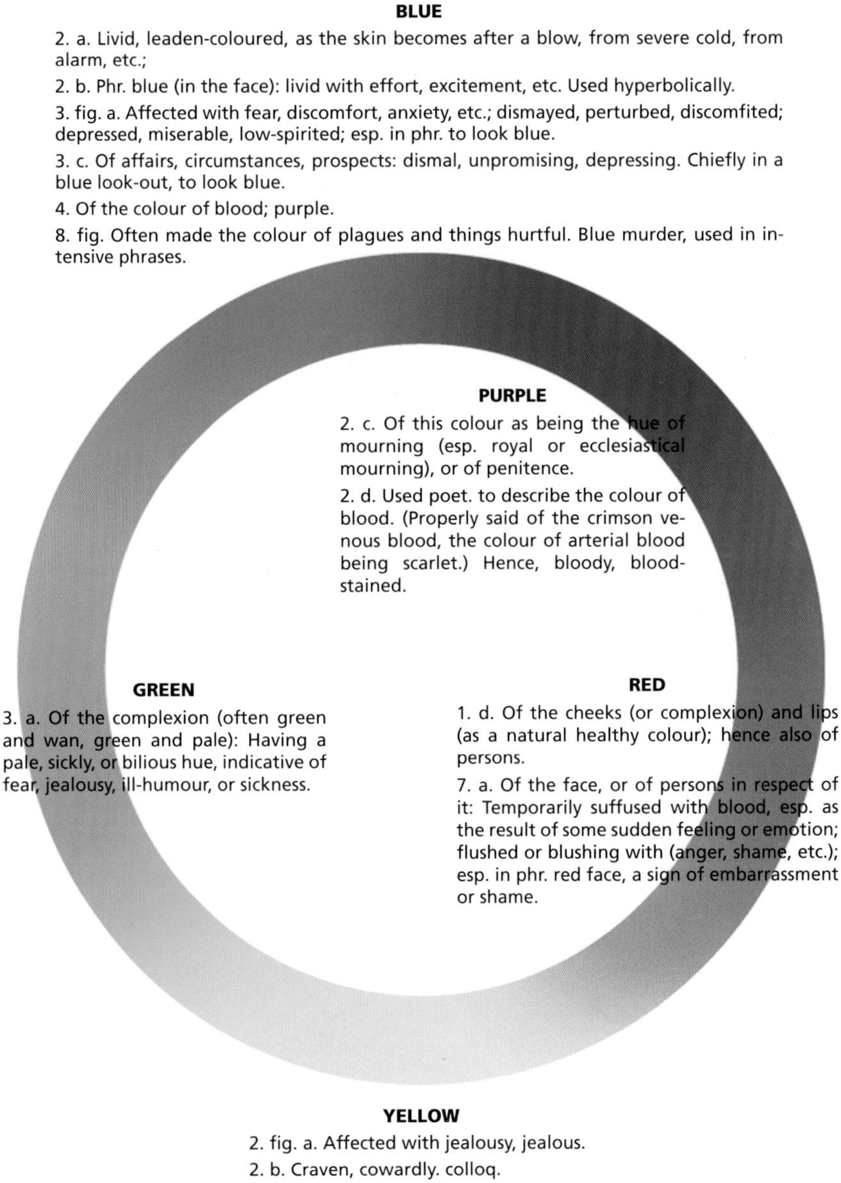

BLUE

2. a. Livid, leaden-coloured, as the skin becomes after a blow, from severe cold, from alarm, etc.;

2. b. Phr. blue (in the face): livid with effort, excitement, etc. Used hyperbolically.

3. fig. a. Affected with fear, discomfort, anxiety, etc.; dismayed, perturbed, discomfited; depressed, miserable, low-spirited; esp. in phr. to look blue.

3. c. Of affairs, circumstances, prospects: dismal, unpromising, depressing. Chiefly in a blue look-out, to look blue.

4. Of the colour of blood; purple.

8. fig. Often made the colour of plagues and things hurtful. Blue murder, used in intensive phrases.

PURPLE

2. c. Of this colour as being the hue of mourning (esp. royal or ecclesiastical mourning), or of penitence.

2. d. Used poet. to describe the colour of blood. (Properly said of the crimson venous blood, the colour of arterial blood being scarlet.) Hence, bloody, blood-stained.

GREEN

3. a. Of the complexion (often green and wan, green and pale): Having a pale, sickly, or bilious hue, indicative of fear, jealousy, ill-humour, or sickness.

RED

1. d. Of the cheeks (or complexion) and lips (as a natural healthy colour); hence also of persons.

7. a. Of the face, or of persons in respect of it: Temporarily suffused with blood, esp. as the result of some sudden feeling or emotion; flushed or blushing with (anger, shame, etc.); esp. in phr. red face, a sign of embarrassment or shame.

YELLOW

2. fig. a. Affected with jealousy, jealous.

2. b. Craven, cowardly. colloq.

FIGURE 5. Oxford English Dictionary *definitions for color terms related to skin, blood, and emotion.*

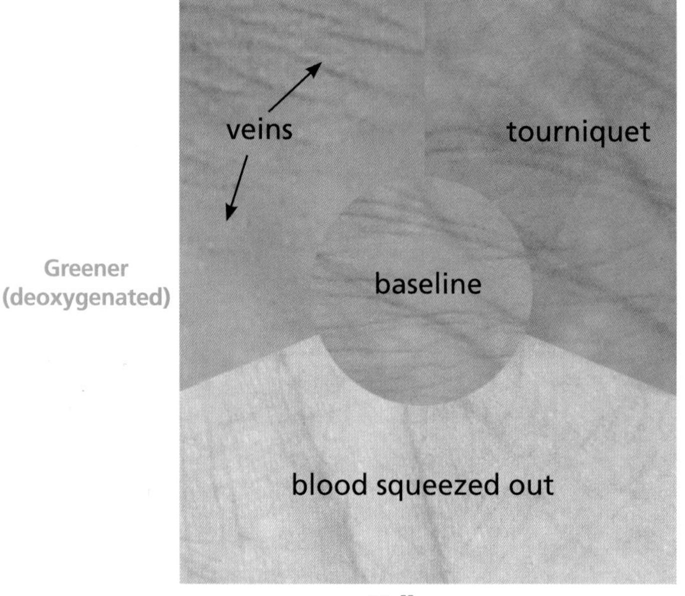

FIGURE 6. *With a tourniquet, relatively oxygenated blood accumulates, which means the skin appears more red and blue, or purple. Skin with veins visible beneath it has a large quantity of deoxygenated blood, meaning the skin appears more green and blue. Skin with the blood squeezed out yellows in relation to the baseline (at the center).*

More Blue	
Physiology	Hemoglobin concentration high
Meanings	Lethargy, choking, blood accumulation (e.g., bruise)
Association	Heavy, sad, strong, cold

More Green-Blue (cyan)	
Physiology	Hemoglobin concentration high, hemoglobin oxygenation low
Meanings	Veins, sickness (cyanosis)
Association	Cold

More Purple	
Physiology	Hemoglobin concentration low, hemoglobin oxygenation low
Meanings	Angry, choking, welts
Association	Slow

More Green	
Physiology	Hemoglobin oxygenation low
Meanings	Sickly, anemic (chlorosis), fear, winded
Association	Weak, cold, happy, clean

More Red	
Physiology	Hemoglobin oxygenation high
Meanings	Embarrassment (blush), exertion (flush), sexual excitement (sexual flesh), engorgement, anger, erythema
Association	Strong, hot, angry, dangerous, aggressive, sexy

More Yellow	
Physiology	Hemoglobin concentration low
Meanings	Fear (blanch), shock, blood loss
Association	Happy, clean, weak, cowardly

FIGURE 7. *Summary of the manner in which skin color changes with blood, what those colors can sometimes mean to observers, and common associations of these colors.*

FIGURE 8. *A case of hypertrichosis, a condition wherein hair grows on skin that is normally bare.*

FIGURE 9. (**a**) *Representative primates without our full color vision. One can see that they're furry.* (**b**) *In most New World Monkeys only the females have color vision.* (**c**) *In most Old World primates like us, both males and females have color vision. Both of these latter groups (i.e., [b] and [c]) tend to have bare spots, as you can see here.*

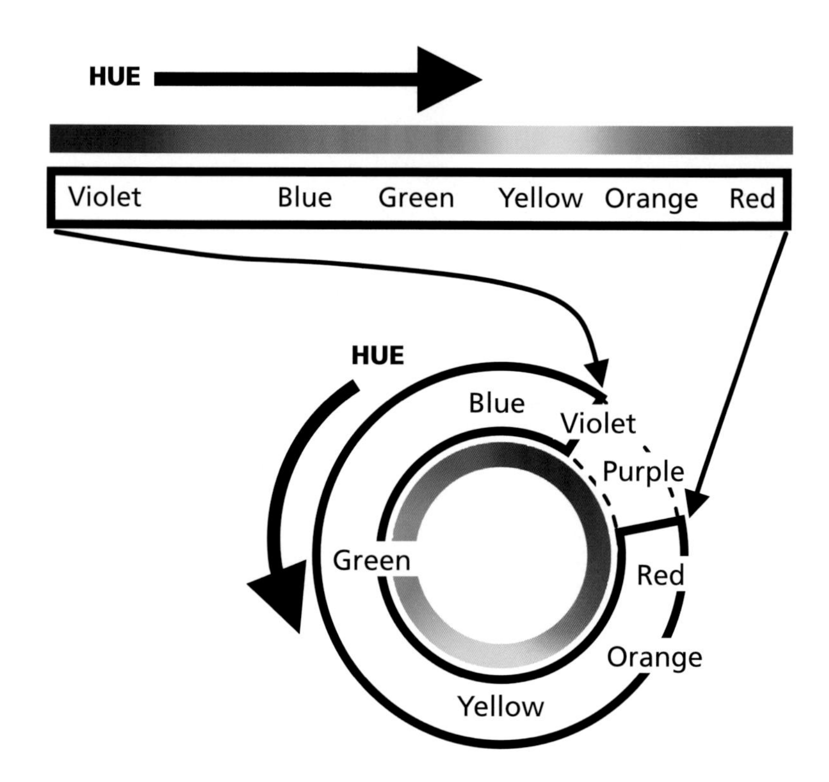

FIGURE 10. *Hues are not well illustrated by looking at a rainbow (top). The rainbow misses the purples and doesn't represent the fact that hues go in a circle, with red changing slowly to violet via passing through purple (bottom).*

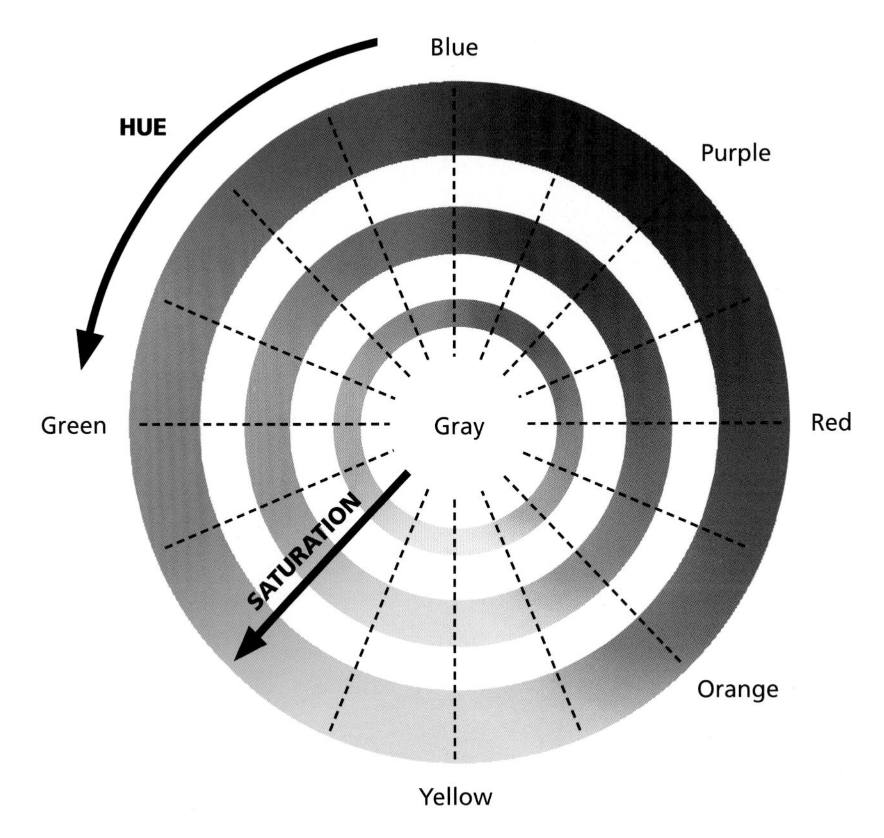

FIGURE 11. *Together, hue and saturation (the two non-brightness dimensions of color) make a flat disk, with hue represented by the location around the circle, and saturation represented by how far a color is from the center of the disk.*

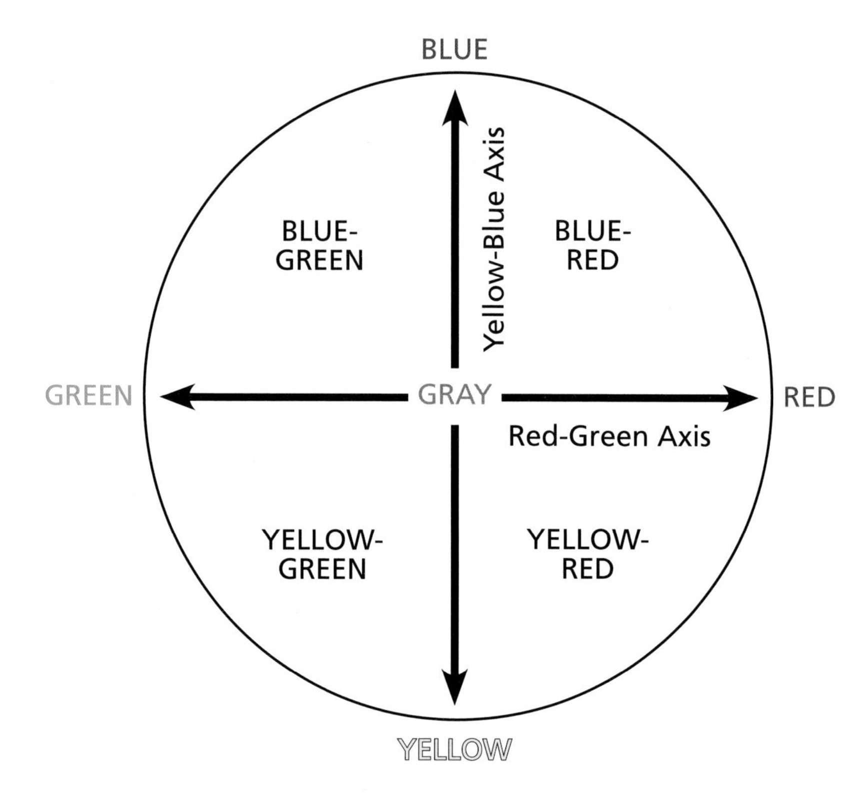

FIGURE 12. *The flat color disk is similar to Figure 11, but no longer illustrates location on the circumference (hue) and distance from the center (saturation). Instead, it shows a vertical yellow-blue axis, and a horizontal red-green axis. This helps illustrate the perceptual opponencies in our color vision: that opposite sides of the disk are perceptually color opposites. Red-green changes are experienced by our brains thanks to comparisons between our long- and medium-wavelength sensitive cones, L and M, respectively: greater activity in L cones relative to M cones leads to redder perceptions; greater activity in M cones relative to L cones leads to greener perceptions. The blue-yellow changes are experienced by our brains thanks to comparisons between our short-wavelength sensitive cones, S, and the average of the other two: higher responses from S cones relative to the other two lead to bluer perceptions; lower responses to S cones relative to the other two lead to yellower perceptions.*

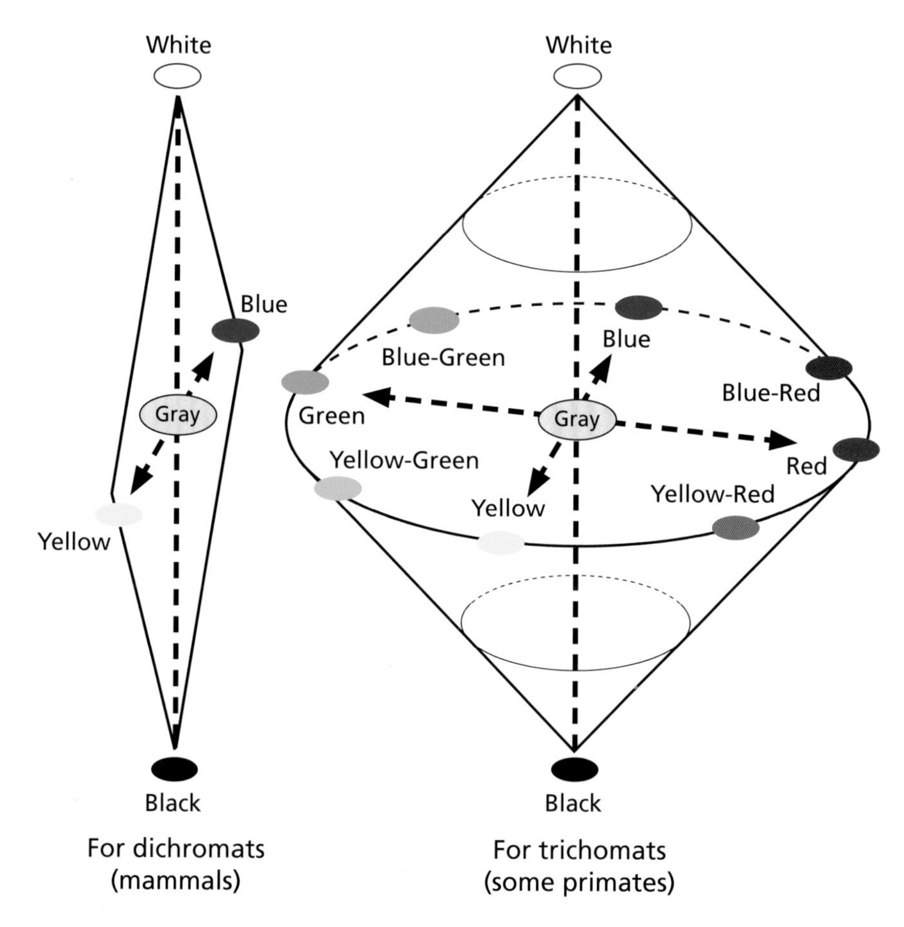

FIGURE 13. *Without the extra cone that trichromats (right) have, most mammals (left) lose one dimension of color, namely the red-green axis. They are left with a brightness dimension and a single, chromatic dimension extending from yellow through gray to blue.*

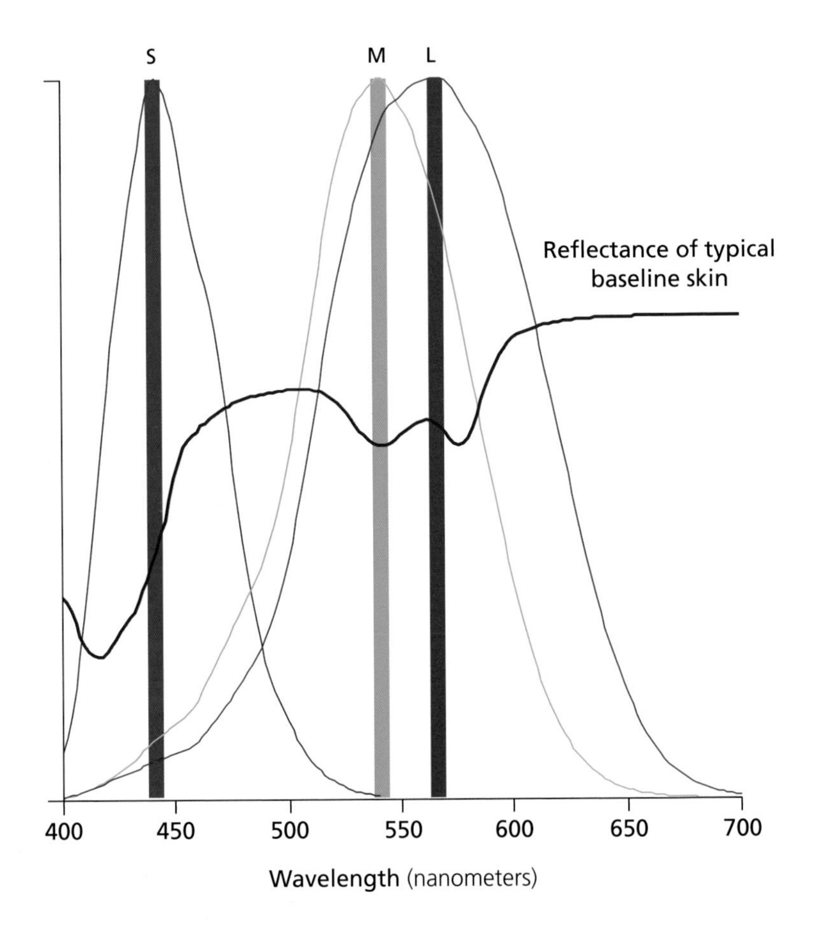

FIGURE 14. *The wavelength sensitivities of our three color cone types, S, M, and L. One can see that M and L have very similar sensitivities (maximum sensitivities are about 535nm and 562nm, respectively). Also shown is a model spectrum from human skin. A characteristic feature of skin is the "W" shape formed at about 550nm. Note how the left trough of the "W" and the peak of the "W" are near the maximum sensitivities of the M and L cones, respectively. The "W" shape occurs due to oxygenated hemoglobin in the skin, and having the M and L cone sensitivities fall where they do is what gives us the ability to read minute changes in skin color so easily.*

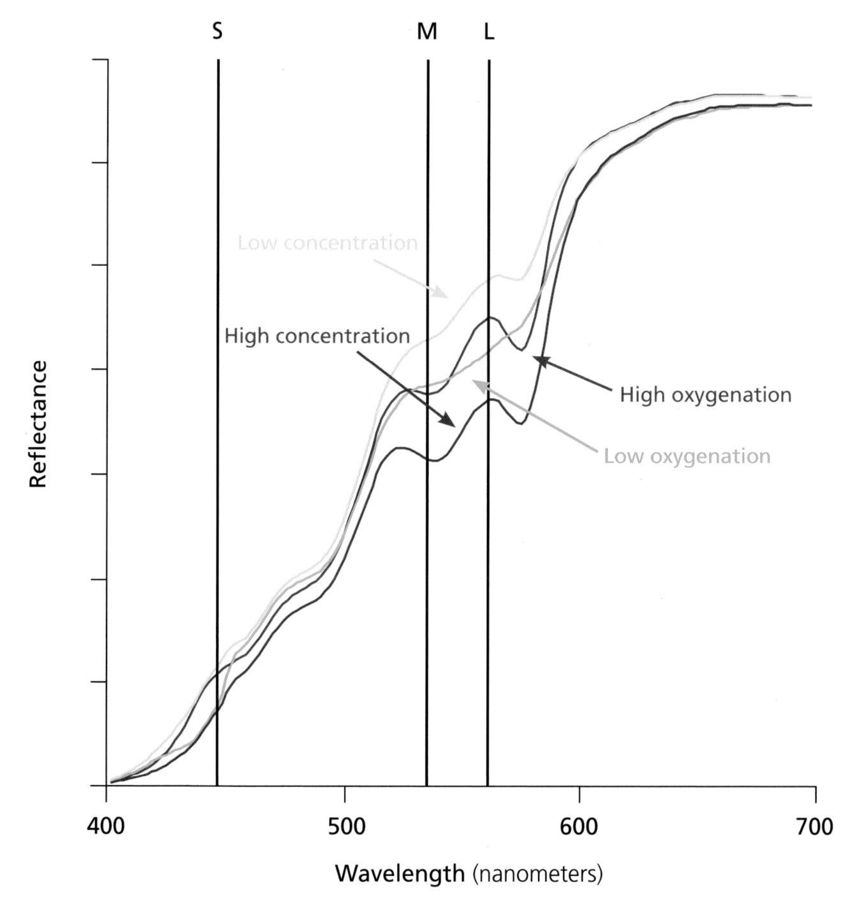

FIGURE 15. *The spectrum of skin as seen by the retina (i.e., after having been filtered by the eye) varies depending on the underlying blood. The blue and yellow curves show skin when blood (hemoglobin) concentration is high and low, respectively. Greater blood concentration drops the location of the "W," and lower concentration raises it. The red and green curves show skin when blood oxygenation is high and low, respectively. The only region of the spectrum that varies is around the "W" shape. By computing the difference between the activation of L cones and M cones, it is possible to sense the oxygenation of the blood. Note that the overall height of the "W" shape doesn't much change as oxygenation changes, meaning there will be relatively little yellow-blue color change with oxygenation. By computing the average activation of the M cones and L cones and comparing this average to the activation of the S cones, our brain can determine the concentration of blood in the skin.*

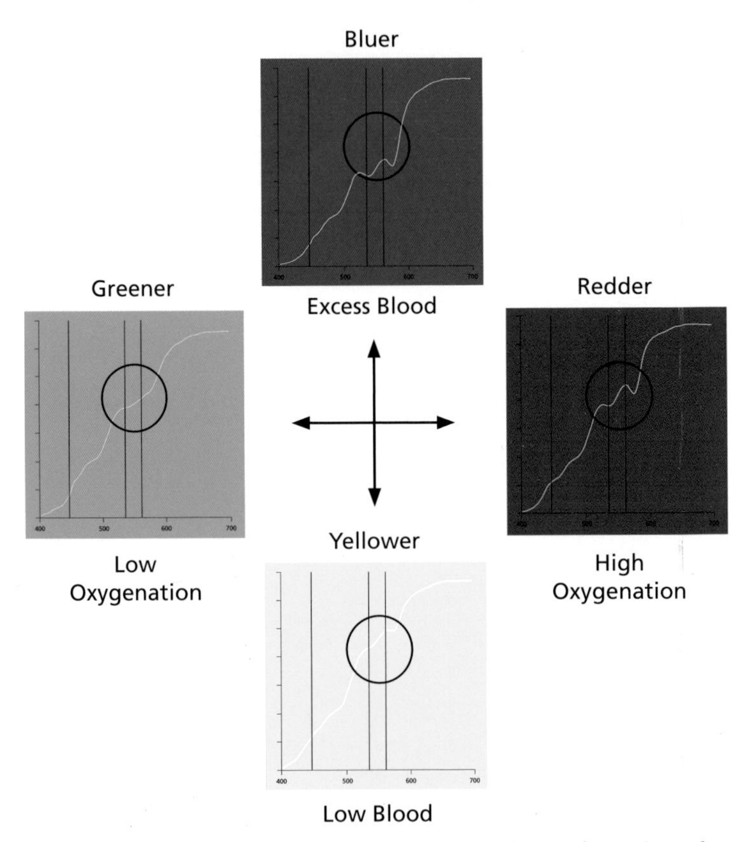

FIGURE 16. *Each of the plots shows the spectrum of light from skin after it has gone through the eye (i.e., just before it reaches the cones). By examining the blue and yellow plots, one can see how the spectrum changes as the concentration of blood changes. The main change is in the 550nm region, where the "W" shape is located, and the entire skin spectrum there lowers as blood accumulates. The result is that blood in the skin is seen as blue because the M and L cones together are activated less relative to the S cone. The red and green plots help us grasp how the spectrum changes as the oxygenation of the blood in the skin changes. The main change also occurs in that same 550nm region, except this time the difference lies in the shape of the "W": the "W" is more accentuated with greater oxygenation and disappears with greater deoxygenation. Because the center peak of the "W" lies at approximately the maximum sensitive wavelength for the L cone, as the "W" becomes more accentuated, the L cones become more activated in comparison to the M cones, and the skin appears more red. The function of the two color cones allows us to register difference along the red-green dimension (including changes in blood oxygenation) without interfering with our ancestral mammalian ability to register differences along the blue-yellow dimension (which is also useful in noting changes in blood concentration).*

"opponent" axes, one horizontal and the other vertical. There is also a third opponent axis not shown in Figure 12—the white-black axis, representing brightness.

These axes are helpful for making sense of the organization of our perception of color. They are also essential for understanding what occurs in our eyes when we see color. Remember cones, discussed earlier in this chapter? Cones, which are referred to as S, M, and L, are neurons that fire when reached by wavelengths of light that are either short, medium, or long, respectively. Your eyes use these cones to perform three calculations, each having to do with one of the three opponent axes: black-white (brightness), blue-yellow, and red-green. Your perceptions of brightness, or variations from black to white, are roughly due to the sum of the activations from all three cone types (though the M and L cones tend to matter most for this): greater overall activation leads to perceptions of greater brightness. Your perceptions of blue versus yellow are due to the difference between the activation of S cones and that of the average activation of M and L cones: when S activations are higher than the average activations for M and L, you perceive more blue; when S activations are lower, your perceive more yellow. Your perceptions of red versus green are due to the difference between the activations of L cones and M cones: when L activations are higher than M activations, you perceive more red; when L activations are lower, you perceive more green.

We are now in a good position to explain how the experience of color differs between primates with full color vision and most other mammals. The color space for the typical mammal (and those primates without full color vision) is not the one we have thus far been describing. Mammalian color vision is, in fact, missing an entire dimension of color—mammals possess only two cone types. Rather than possessing both the M and L cones, they have just a single cone (called "M/L"). They do not have the opponent mechanism responsible for red-green variation, so don't perceive a red-green axis. It is this red-green axis that is evolutionarily new among color-equipped primates. What is a two-dimensional disk for us (Figure 12) was a mere one-dimensional line for our color-blind ancestors. Our perception of color forms a three-dimensional, double-cone shape, as shown on the right of Figure 13, whereas the typical mammal's color perception forms merely a flat, two-dimensional diamond, as shown on the left in Figure 13. As a result, whereas primates with color vision have infinitely many hues that smoothly change along the circumference of a disk, the (dichromat) mammal has only two hues—yellow and blue, each the opposite of the other. Therefore, the typical mammal possesses a kind of color vision that is significantly less rich than ours, which is why we tend to say mammals lack color vision.

In all fairness, it is not quite true that most mammals lack color vision. After all, mammals do perceive two hues: blue and its opposite, yellow. That blue-yellow dimension along with the brightness dimension effectively creates infinitely many colors (but, again, in only two hues). If we were to insist that mammals' less rich color palette does not count as color vision, then birds, reptiles, and bees could say that *we* don't have color vision, for they possess a color dimension above and beyond what we have! Just as we perceive richer colors than most other mammals, these non-mammals are capable of even richer color perceptions than we are. But don't let any of these non-mammals tell me I don't have color vision.

Blood in the Eye

We now have a better understanding of the basic principles underlying our perception of color, but we still don't know why we perceive certain objects as having certain colors. Why does grass appear green rather than purple? Why is the sky blue, not red? So far, we only know facts such as that our experience of red will be pure, our experience of purple will be of a mix of red and blue, and we will never see hues that are reddish-green. Our color space is, after all, just a color palette, and a color palette can be used to paint objects however you like. But our world is colored using our palette in a particular fashion. Grass is green, not purple. The sky is blue, not red. Which objects in the world get "painted" which colors depends on the specific spectral sensitivities of our visual hardware. It's not enough to know that we have short-, medium-, and long-wavelength sensitive cones with which certain comparisons are made. We need to know what specific wavelengths of light the cones are actually sensitive to. Recall from our earlier discussion, in the section titled "Green Photons," that color is not actually about wavelengths of light. Instead, color is about the perception of the complex distributions of light of all wavelengths (in the visible part of the spectrum) that emanate from each object. We have evolved to perceive the colors not of photons, but of certain *objects and surfaces*—especially surfaces of *skin*.

Figure 14 shows the wavelength sensitivities for each of the three cone types. You'll notice that, strangely, the sensitivities of the M and L cones fall nearly on top of one another. At first glance, this seems like a terrible engineering design. A smarter design would be to sample wavelengths in a uniform fashion—an S cone that samples the short wavelengths, an M cone that samples the middle wavelengths, and an L cone that samples the long wavelengths. That's what cameras do. That's what birds, reptiles, fish, and bees do (although they sample with *four* cones). It seems like a waste to sample nearly the same wavelengths of light with two different cone types.

But there's a method to the madness, and a hint of it can be seen in Figure 14, where—along with the cone sensitivities—you can see what the usual spectrum of human skin looks like. The "W" feature shape in the middle is particularly important; that little wiggle is actually due to the way that light is absorbed by oxygenated hemoglobin in the blood under the skin. Notice how the maximum wavelength sensitivities for the M and L cone types match the left trough and the center peak of the "W," respectively. *That* synergy turns out to be crucial to our empathic ability.

Figure 15 shows how the spectrum of skin changes as the concentration and oxygenation of the blood under the skin changes. (This figure shows the spectrum of skin *after* the light has gone through the eye, just as it hits the cones, whereas Figure 14 showed skin's spectrum before it hits the eye. The eye is not a perfectly transparent medium, and so some of the light is absorbed before it reaches the retina.) The blue and yellow curves show how the spectrum of skin changes with the concentration of blood. The main change that occurs when skin shifts from having a low concentration of blood (yellow) to having an excess concentration of blood (blue) is that the spectrum of the skin in the "W" region lowers; there is little change anywhere else in the plots. The average activation of both the M cones and the L cones lowers as the concentration of blood increases, and the skin appears more blue. When the opposite occurs, the skin appears more yellow. (Greater blood concentration also lowers overall brightness, whereas lower blood concentration increases skin's brightness, and one wonders whether this difference could underlie why blue is considered a "dark" color and yellow a "light" color.)

The red and green curves in Figure 15 show how the spectrum of skin changes as blood oxygenation changes. As you can see, the overall height of the "W" shape doesn't change. The change here is much more subtle: as the oxygenation level shifts from low to high, the "W" shape becomes markedly more visible. Because the M cones are the most sensitive to the left trough of the "W" and the L cones are the most sensitive to the central peak of the "W," the more the shape of the "W" is accentuated by oxygenation, the more the L cones' activation rises relative to the M cones'. This leads to a greater perception of red in the skin. Similarly, deoxygenation causes skin to appear green. Figure 16 shows the same four curves as in Figure 15, but positioned around the color disk from previous figures.

It is this closeness of M- and L-cone wavelength sensitivities that allows us to see color changes in the skin as the blood changes in these two ways. Importantly, note how similar the four skin spectra look in Figure 15, especially the spectra showing deoxygenated versus oxygenated blood. Being able to notice such a small variation in color is not easy, and the wavelength sensitivities of our M and L cones are just where they need to be to see this—namely, at the left trough and central peak of the "W" feature, respectively. This placement not only allows us to see the red-green dimension, but also lets the M and L cones act like the ancestral single M/L cone, which allows us to still see the blue-yellow dimension. Since our new ability to sense oxygenation changes in skin via our new red-green perception does not interfere with our ancestral ability for blue-yellow perception, we can infer that we've been selected to maintain it. Furthermore, the precise location of the M and L cone sensitivities are what allow us to distinguish blood concentration changes from oxygenation changes so clearly. If the M and L cone sensitivities were located instead at the central peak and the *right* trough of the "W" shape, we would still perceive changes in red-green perception, and

M and L would probably be sufficiently similar to the ancestral M/L cone to enable our blue-yellow vision to function, but changes in blood concentration would affect red-green perception much more strongly than it does now. Red-green changes would be more ambiguous; it would be harder to tell whether skin color changes were due to blood concentration or to oxygenation changes. Our real M and L cone sensitivities help keep our perceptions of these blood changes more independent of one another.

Only the red-green dimension of color is new; the blue-yellow dimension has been in use for tens of millions of years, since before we had bare skin. Does this mean that only red-green variations on the skin have emotional significance? Almost certainly not. Color signaling could have evolved by tapping into *any* colors we could see. Because blood concentration and blood oxygenation modulations are approximately independent, color signaling was able to pull from both the ancestral blue-yellow dimension *and* the newer red-green dimension, creating color signals that are all the more dazzling to the eye.

Splash of Color

Color is broadly about our perception of the distributions of wavelengths of light emanating from objects. But our eyes are not spectrometers: our eyes are not capable of measuring how much light of each wavelength is coming into the eye. We would need tens or hundreds of cone types to hope to measure that, rather than just the two, three, or four cone types that animals typically have (although some crustaceans have more). Instead, we use our limited number of cones to sample wavelengths that are of the greatest use to us. If "X" is the most important thing—e.g., skin—then cone sensitivities will be selected that best allow the animal to see X. The animal's color vision will be designed for X.

Crucially, notice that these colors out in the world are a kind of useful fiction. It is as if evolution took crayons and labeled with specific colors the things that are important for our survival and reproduction. Yet the real world has no such labels. Donald D. Hoffman, a professor of cognitive science at the University of California at Irvine, likes to emphasize this point, but instead of a crayon metaphor, he prefers to use a computer desktop metaphor. In a computer desktop environment there are objects, or icons, of various colors and shapes that are displayed on a simulated desk and can be moved, opened, and even thrown into a trash can. This computer desktop is a visual representation of the internal workings of the computer. But how was this particular representation chosen? Not via evolution in the natural world, of course, but via market pressure on engineers over the last couple decades to design computer desktop environments—i.e., visual systems for seeing "inside" the computer—that are better suited for the human brain. Computer desktops have evolved to enhance human interaction with computers, analogous to how human visual perception has evolved to enhance human interaction with the world.

The nice thing about this analogy is that no one believes that computer desktops actually represent any of the computer's *true* properties. We all understand that the icon's position on the screen, its color, its size, the manner of its manipulation, and so on are "illusions," in the sense that they exist merely to help you interact and navigate within your computer. They are useful fictions. But if engineers chose the useful fictions of desktop icons to optimize human-computer interaction, might it be the case that useful fictions are also what evolution chose to optimize human-world interaction? That is the point behind Hoffman's analogy. Evolution does not care if the visual system actually provides a true representation of the world around it, so long as the visual system generates perceptions that allow the animal to function successfully and spread its genes more quickly than its competitors. Evolution only selects for visual systems that propagate; accuracy is secondary. Dr. Hoffman believes that, on the basis of these observations, we can no longer be sure of any of our beliefs about the external world. I think Hoffman goes off the deep end when he takes these kinds of skeptical arguments too much to heart (and I admit, I was worried he would no longer pay for my lunch if he thought I didn't exist), but the moral we get from the desktop-world analogy is a great one, and will come up again in the forthcoming chapters.

As we mentioned above, an animal's color vision will evolve to be useful for seeing whatever is of greatest use, or "X." But what about all the other things in the world besides X? In our case, what about objects other than skin? Once evolution has created a color palette for X, the colors in that palette end up being seen on everything else, too. Our red for oxygenated skin ends up splashed on sunsets, rubies, and ladybugs. These latter cases of red are capricious accidents; there was no evolutionary pressure for the spectra of sunsets, rubies, and ladybugs to be categorized as similar in color. In reality, the three differ significantly from one another—just not as far as our cones can detect. We, and other animals with color vision, end up seeing the entire world through X-colored glasses. In our case, there are red and green variations all over the natural world, most of which are not meaningful. Such red-green perceptions are not just fictions, but probably *useless* fictions. That is, when seen on skin, red-green perceptions are useful, but seeing red flowers and green leaves provides no benefit. It is as if special skin spectacles have been sewn over our eyes, and although they help us see the inner lives of the people around us, they can mislead us about the non-social world (e.g., we may mistakenly believe there is something objectively similar about sunsets, rubies, and ladybugs). This is a telling reminder that the way we see is designed to help us reproduce, whether or not it leads to more objective perceptions of the world around us. (Perceptions of blue and yellow, however, must be useful in some other way in nature, because mammals without bare skin have this dimension of color too—although what exactly it is for is unclear.)

Our color view of the world is a strange one, then. Animals without our number or types of cones see the world as colored differently. Our colors are splashed here and there (except on skin) in a way that makes little sense, had it been done on purpose, and our perceptual view of the world is much less objective than we typically realize. In fact, animals that lack color vision altogether have, in a sense, a more objective—a less colored—view of the world. Colors don't end up thrown onto objects there's no practical reason for them to be on...but that objectivity is gained at the cost of not receiving any wavelength information at all. Animals without color vision can only sense how much light, overall, reaches their eye. Thinking in this way has given me a new respect for black-and-white photography. I used to feel it was an inferior medium, and couldn't understand the fascination some people have with it. Now, however, I suspect black-and-white photographs show their subjects closer to how they "really" look. Bees, birds, and humans could all approximately agree about the look of a black-and-white photograph, for they'd all see approximately the same thing. But they'd all see different colors in a color photograph. If you were to send a photograph into space for aliens to one day discover, a black-and-white photograph would be much easier for them to make sense of. If you sent a color photograph, then even if the aliens saw in color, their color vision wouldn't be based on *our* peculiar hemoglobin. What they saw in the photograph would be colored quite differently.

In Living Color

Knowing that color vision is meant for seeing the subtle and not-so-subtle changes in our skin color probably won't increase sales for "flesh"-toned wall paint. But I hope you now see that your uncategorizably colored skin is actually wonderfully colorful, because our eyes have evolved to turn skin into a full-color monitor on which we are able to watch emotional dramas replete with sex and violence. To animals without color vision, or without *our* kind of color vision, these skin-shows are invisible; they may even wonder whether our ability to see these deeply personal signals means we're magical in some way. We're not magical; we're just empaths. Real, live empaths.

CHAPTER 2
X-Ray Vision

"Now watch," said the Zebra and the Giraffe. "This is the way it's done. One—two—three! And where's your breakfast?"

Leopard stared, and Ethiopian stared, but all they could see were stripy shadows and blotched shadows in the forest, but never a sign of Zebra and Giraffe. They had just walked off and hidden themselves in the shadowy forest.

—RUDYARD KIPLING,
"How the Leopard Got His Spots" (*Just So Stories*)

Secret Chimp

Imagine that you're suddenly transformed into a squirrel. After getting over the fur, your two-pound svelte frame, and the shock that your buddy's "magic wand" really *was* magic, you notice that your television and couch are missing. In fact, you're in a tree in a rainforest in Uganda. And you're not alone. You're with others of your (new) kind, and they seem to be agitated by something; they aren't in the mood to greet you. They're frantically peering down below, and although you can't see a thing, you have a bad feeling. This is *not* the day to be transformed into a rodent. And then you hear it: hooting and screeching below you from every direction. You know that sound: chimpanzees.

Phew! You were worried for a moment there. But there's no snake, no palm civet, no leopard. Just good-natured chimpanzees—the kind that look hilarious when dressed in suits on television commercials. Now, how are you to get to the embassy and explain why you don't have your passport with you?

As you begin climbing down, amazed at your claws, you wonder why the other squirrels have suddenly disappeared. What *are* they so afraid of, anyhow? Suddenly your face turns ashen white (or it would if you were still human) as you remember something you saw on TV. Chimpanzees sometimes hunt mammals like smaller monkeys...and squirrels! Everything you learned from the Kratt Brothers television show comes flooding back: Chris and Martin Kratt had followed chimps during a hunt, filming them as they caught the little squirts, bit their little heads clean in half, and ripped their limbs off. "*I'm* a little squirt!" you yell out, but with your new vocal tract you only manage to say, "Kwurp, kwurp, kwurp!", something that, judging by the rapidly increasing closeness of their calls, the chimpanzees apparently understand to mean, "Squirrel over here, cooked rare!" (See Figure 1).

You dash through the canopy, bounding deftly from branch to branch and tree to tree, avoiding the sounds of pursuit. You note how lucky you are to have good hearing, since you've never actually *seen* any of the chimp pursuers on your tail. In fact, you're having trouble seeing much of anything at all. Layer after layer of leaves, many of

a

b

FIGURE 1. (a) *You after your wizard friend transforms you into a squirrel.* (b) *A chimpanzee wondering whether you're on the menu.*

which are bigger than your now-diminutive, chimpanzee-bite-sized head, block your view. Even if you'd known where you were before, you're lost now. Luckily, you're faster and nimbler than the chimpanzees, which helps counter the fact that they are about fifty times your weight. You can also move very quietly, especially in comparison to the racket those chimps are making. But if the chimps can't hear you, how are they following you so doggedly? It's as if those chimps can see you, even though you can't see them. Do they have secret chimp X-ray vision goggles that let them see through these seemingly opaque leaves? Just as you reassure yourself that this is crazy-talk, you're caught—and never get a chance to follow up on the idea.

This chapter is about these secret chimp powers, a kind of "X-ray vision" they have. You and I have these powers as well, making us a terror in the woods. The story of this superpower starts with cyclopses, and why there are so few of them.

Why No Cyclopses?

There aren't many cyclopses in nature, and those that exist don't live up to expectation—getting these animals tipsy and stabbing them through the eye with a stake would be much less impressive than when Odysseus did it. Real "cyclopses" tend to be crustaceans like water fleas (Figure 2a) or invertebrates like our early fish-like ancestors the lancelets. There are no vertebrate cyclopses. Some fish, frogs, and lizards, and their distant relatives, have a third eye (a "parietal eye") directly on the tops of their heads (Figure 2b), which is similar to being a cyclops in that the parietal eye is not paired with another eye. However, although these eyes are light-sensitive, they are not really eyes at all because they are not capable of forming an image of the outside world. Rather, they seem to be involved in thermoregulation. The closest thing to a cyclops in the higher vertebrates is an anencephalic infant. But that's nature's *mistake*, not intentional design.

One reason for the scarcity of cyclopses is that they can't see what's behind them. Most of the Earth's animals with vision *can* see much of what's behind them, presumably because they need to. They accomplish this by having two eyes that face opposite directions. This is true of squid, insects, fish, reptiles, dinosaurs, birds, and many mam-

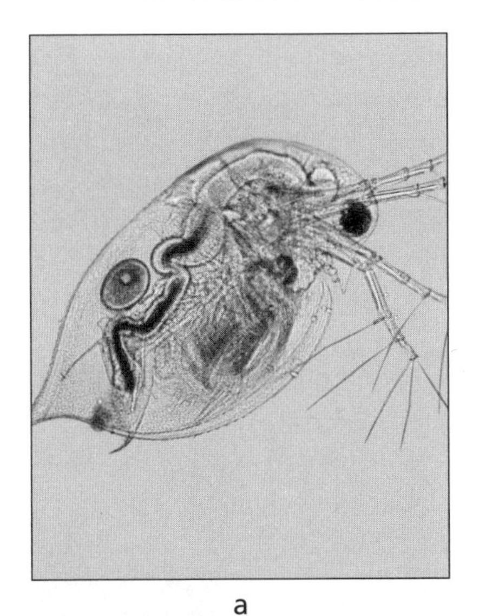

a

b

c

FIGURE 2. *Cyclopses.* (**a**) *Single eye of a water flea.* (**b**) *Parietal eye on a lizard.* (**c**) *We only find mammalian cyclopses, like this mythical roman god, in fiction and legend.*

mals, like rabbits and horses (Figure 3). You can't sneak up on these animals from behind. But if panoramic vision—i.e., having a view of the front and rear—explains why there are so few cyclopses, then what about us? We have two eyes, but they both point in the *same* direction. There must be something we gain from this—something beneficial enough to overcome the disadvantage of complete blindness to what's behind us. I'll get to that later, for that is the central

FIGURE 3. *Most animals have two sideways-facing eyes, giving them a panoramic view of nearly everything that's around them.*

question of this chapter: "Why do our eyes face forward?"

For the moment, however, let's turn our attention to something so obvious about the animals in Figure 3 that one is apt to overlook it. These animals have eyes facing opposite directions, as I mentioned, but more specifically, they have eyes that are on the *sides* of their heads, like in Figure 4a. This is not the only possible way of having panoramic vision. A single eye could be placed on the very front of the head and the other on the very back, as illustrated in Figure 4b. You never find this in nature (and not even in fiction, as far as I know). Why not?

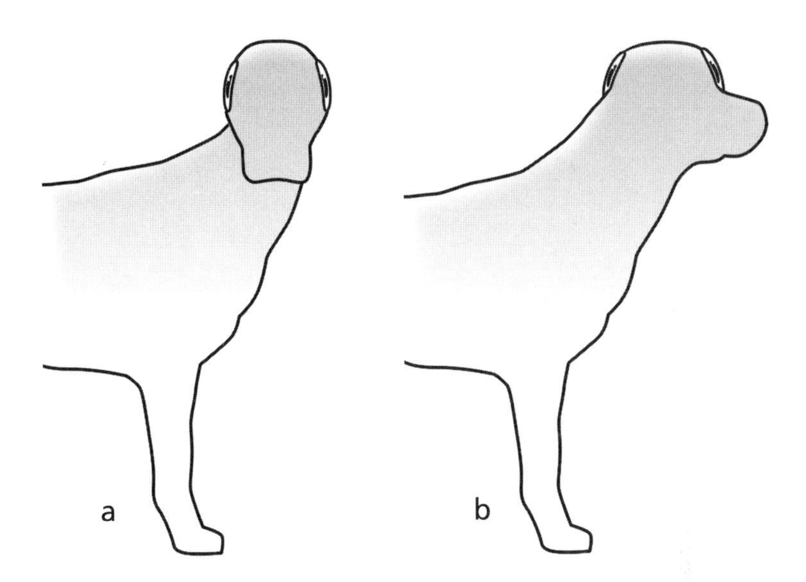

FIGURE 4. (**a**) *Lots of animals have sideways-facing eyes, like this imaginary animal.* (**b**) *But no animal has one eye in front and the other in the back. Why not?*

The reason has to do with the fact that the sideways-facing eyes in Figure 3 don't just see the opposite sides of the animal. Instead, they typically have some overlap, i.e., parts of the world that both eyes can see, called the binocular region. Seeing a part of the world through two eyes has certain advantages, advantages that are best employed in front of you. For animals with sideways-facing eyes, the binocular region is located in front of them (Figure 5a), but also above and behind (Figure 5b). For the animal in Figure 4a, *you* are in its binocular field because you can tell that each of its eyes can see you. For an animal with an eye on the front of its head and another on the back, however, any overlapping regions of its visual field would be along the *sides* of the animal instead of the front, and there's less advantage in having your most powerful vision directed to the sides. (Our vision is a variant of having eyes on the sides of the head, in the sense that our eyes are placed on the left and right of our midline, so that our binocular region is also focused in front.)

What are these advantages of binocular vision? The typical answer someone knowledgeable about vision would give is stereo vision, or

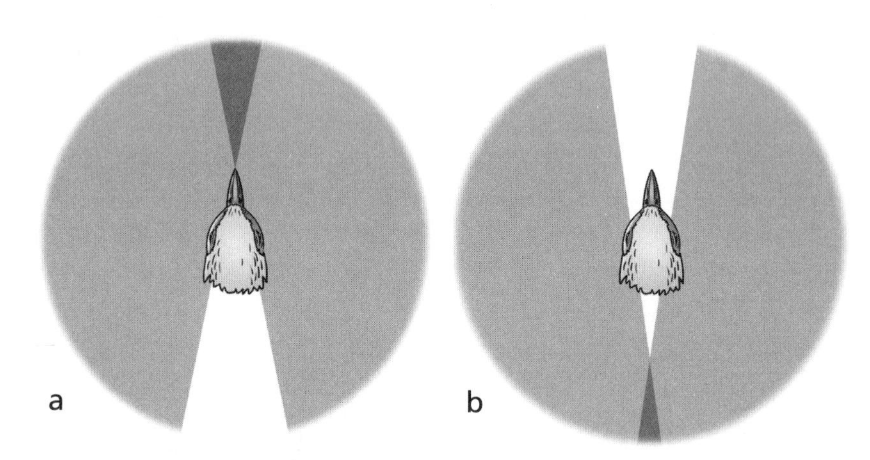

FIGURE 5. (**a**) *The regions visible through each eye for an animal with sideways-facing eyes (here, a bird). There is a region in front that both eyes can see, the binocular region. Also, each eye can see the beak, and although the beak blocks some of the forward view of each eye, the other eye covers what is missed, meaning nothing is missed out front. That is, the animal can simultaneously see its beak and see everything beyond it. (Note that there is also a region nearer to the animal, before the binocular region, which is a blind region; neither eye can see that region in front of the eyes. No visual feedback is possible here, but it is a safe place to put bodily appendages because they can't block the view.)* (**b**) *The figure in (a) is for when the bird is looking forward, but if the bird looks backward, the visual field of each eye shifts, and a new binocular region appears directly behind it. Some but not all animals with sideways-facing eyes can do this (e.g., rabbits).*

the ability to see depth. The study of binocular vision is so historically wrapped up with stereo vision that courses and books about binocular vision are primarily devoted to depth perception. However, through my research I have come to appreciate a power of binocular vision no one has yet noticed: the power to see *through* things. Understanding this "X-ray power" is crucial for understanding why we have forward-facing eyes.

Before discussing our X-ray power in general, I want to talk about a special, very fundamental kind of X-ray vision we get from binocular vision, one that helps further drive home the advantages of panoramic vision from sideways-facing eyes rather than front-and-back-facing eyes: the power to see through yourself.

I Can See Right Through Me

Before an animal can worry about how to most *ably* see what's out there in the world, it first has to make sure it can see what's out there at all. An eyeball placed on the inside roof of an animal's mouth would not be of much use. But where *should* the eyes be placed? Most spots would be better than the roof of the mouth, but bodies also have gangly parts that could possibly block an eye's view. And, in fact, one typically *wants* some of those body parts to be out in front of the eye, body parts like a mouth, a muzzle, a nose, whiskers, or hands that are designed to physically interact with things out in front. Animals want these appendages out in front of the eye not only because these appendages are good at interacting with the world and eyes are not, but also because it is useful to see their own appendages so they can help guide these appendages' interactions with the world. And that leads to a riddle that animals with vision had to solve. Namely, how can they put appendages out in front of their eyes, yet not occlude their view of the world?

This is a riddle that video game makers have had to grapple with as well. How do you let the player see his own character without obstructing the view of the game? If you play a game in first-person perspective mode, your character holds a weapon that covers a significant portion of the screen (e.g., see Figure 6a). If you play in third-person perspective mode—where you see the entire character from behind—then there are also significant parts of the view missing because the character's whole body blocks part of the view (e.g., see Figure 6b). That's why many games allow the player's viewpoint to roam around the character rather than remain locked into place. Allowing our eyes to float around our bodies and relay information to our brains would be one potential solution to the problem of seeing past our own appendages.

Nature found a different solution, however, and you can see it for yourself by looking at your own nose. If you close one eye and wiggle your nose, you'll see it in the bottom corner, blocking the open eye's view of whatever lies beyond it, as illustrated in Figures 7a and 7b. But if you then open the other eye, you'll see what's behind your nose via this other eye (Figure 7c). You'll still perceive your nose to be there, but only as a transparent shape through which you can see

FIGURE 6. *Two simulated video game screenshots.* (***a***) *A case of first-person perspective, where one's arm and gun occlude some of the scene.* (***b***) *A case of third-person perspective, where the character's body blocks some of the scene.*

the scene beyond. Each eye sees the nose, and each eye has its view blocked by the nose. But the views blocked by the nose are different for each eye, so the pair of eyes together doesn't miss a thing—*including* your nose. If, instead, your eyes were on the front and back of your head like in Figure 4b, then any appendages out in front of the front eye would simply occlude its view.

Having eyes on the sides of the head (or on either side of the midline, like ours) rather than on the front and back, then, is not only useful for placing the powerful binocular region in front of us so we can better see what we're looking at. Having eyes on the sides of the head is also crucial for seeing past our bodily appendages. And not just seeing past our own appendages, but simultaneously seeing them as well, thereby gaining the advantages that come with having this visual feedback. This may be one of the most fundamental reasons why animals have eyes on either side of their head. The resulting binocular vision gives us a very basic kind of X-ray vision—the topic of this chapter. Of course, at the moment we're talking about the X-ray power of seeing through one's own body, not the world. We'll get to the world later.

Being Here but Seeing From There

We've just seen why having eyes on either side of your head is better than having them on the front and back, and certainly why it is better than having them inside your mouth. But having eyes inside our "bodies" is something we silly humans sometimes do. When you drive, your car becomes an extension of your body, which means your eyes are then located inside your new body. And unless the car's nose—i.e., the hood and engine—happens to be made of glass, some of your view of what's in front is blocked, which is the source of many a scraped bumper.

The fronts of our cars are not supposed to "interact" with the world, and so you can usually get by without seeing your car's nose or seeing through it. Seeing and seeing through front "appendages" is much more important for a tractor operator, however; their appendage is now a shovel or scoop. Some tractors partially alleviate this problem by placing the operator—i.e., the eyes—to one side of

a. What your left eye sees

b. What your right eye sees

c. What you see

FIGURE 7. (a) *Your nose occludes the right part of your left eye's visual field.* (b) *Your nose also occludes the left part of your right eye's visual field.* (c) *But notice in (a) and (b) that what is blocked by the nose in one eye is seen by the other eye. When both eyes are open, we perceive everything, and also perceive the views of our nose, which are now on the sides of our visual fields. We perceive through our solid nose as if it were transparent. The region in the center—where the nose never blocks our view—is the region both eyes see, and is thus the binocular region.*

the shovel arm, so that the operator can see at least one side of the arm and part of what's directly in front of it. The biological solution, though, is illustrated in Figure 8, where the operator's eyes are placed at the tips of long stalks on either side of the tractor. Because giant snail-people tend not to apply for jobs in construction, it is easier to give human operators special goggles to wear, fed with camera views from either side of the tractor. This is something that, as far as I know, is not done for tractor operators, but *is* echoed in the placement of cars' side-view mirrors. As the name suggests, these mirrors are located on the sides of the car, so that (1) each includes a view of the car itself as well as the scene beyond (and behind), and (2) part of the scene is seen by both mirrors (the "binocular region" behind the car). That is, side-view mirrors are placed on a car similarly to how eyes are placed on an animal's body.

There is still a key difference, however. While driving and utilizing your side-view "eyes," you're dealing with two separate images. In fact, you even have to move your head back and forth to see them. But for fun, let's imagine you can see both simultaneously, side by side, in your visual field, perhaps via a dashboard computer display. If our vision of the world were like *this*, then it would be like seeing the pair of images in Figures 7a and 7b side by side; our brain would have to monitor both images, akin to a security officer monitoring two security camera feeds. But that's not how we see the world. Instead, we view it like in Figure 7c: all at once, as a *single* view. There is presumably some advantage to this; perhaps it minimizes the amount of neural tissue required to represent what we see, or minimizes the time necessary for the brain to process the information.

Advantageous or not, it's weird. We have two eyes, with two views of the world, from two different vantage points. How can our resultant perception be as if from only one eye, with one view of the world, from a single vantage point? And where exactly *is* this vantage point, if it is not one of the two *actual* vantage points? That's easy to figure out—we can just open our eyes and ask ourselves where we appear to be seeing from. The answer is right between the eyes, slightly set back behind the upper regions of the nose.

We perceive the world, then, from a vantage point of the middle of

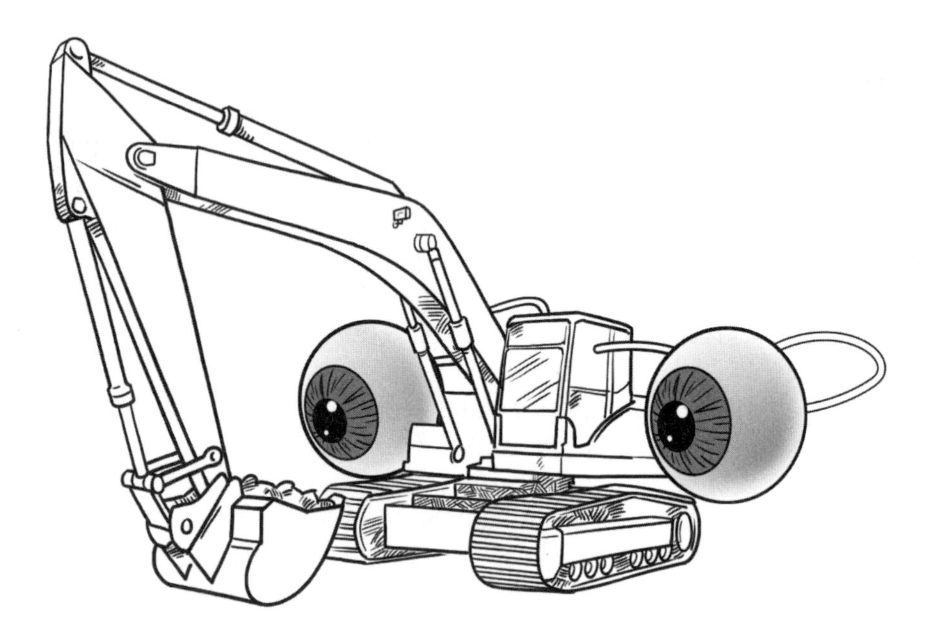

FIGURE 8. *Trying to see what the shovel is doing from inside the cab of a tractor is like viewing the world with a single eye inside your mouth. Better if the tractor operator had long eye stalks reaching out of the cab to see either side of the shovel, to help coordinated digging.*

our forehead, a point at which we have no literal eye. But that's not all. Your view of your nose from that single vantage point should be from above, yet your single, unified perception includes two images of your nose: the right-side view is placed on the left side of your perception, and vice versa, as in Figure 7c. And, even more confusing, the two nose-images are transparent!

How did we end up with perceptions that are so astoundingly false? How do we possibly manage? Luckily, few of us are confused about where our eyes are situated, the number of noses we possess, the location of our nose or noses, or how transparent it is/they are. This brings us back to the computer desktop analogy of visual perception we discussed in chapter 1 (see the section titled "Splash of Color"). Recall that a computer desktop looks as it does because its appearance is useful, not because it accurately represents the computer's innards. Similarly, our visual perceptions have evolved because they're useful, not because they truthfully represent the world in front of our eyes. (Or, when they *do* accurately represent what's out there, it is *because* that representation is useful.) If our brains were content to deal with two separate images, one from each eye, then no fictions would be needed. But if our brains prefer to have a single unified image where there isn't in fact one to begin with, then it has made the decision to "go fictional." Squooshing together two true images into a single one does not, after all, lead to a true resultant image. Therefore, our perception is, in the ways just mentioned, fiction. A fiction we know very well how to interpret. And it is a *useful* fiction because it provides a single unified view of the world, one that includes both the animal's appendages and the world beyond it.

Back to the tractor operator whose goggles show him images from cameras mounted on either side of the tractor's digger arm. The operator is not simply looking at two separate images, one from the left camera and one from the right. Instead, the operator's visual system's ability to create a useful fiction out of the disparate inputs would kick in, and the operator would perceive a single unified perception of the digger arm and what's behind it. Furthermore, because the view from the two cameras is exactly the kind of eye-inputs the visual system is designed for, the tractor operator will need little or no training to get used to it. Sure, he'll perceive two images of the digger-arm, but this is just the kind of fiction the brain knows how to read.

Now let's imagine using this same tractor-operator trick on ourselves. Suppose two cameras were rigged to hover behind and on either side of you, and that they fed the two images into binocular goggles you were wearing (see Figure 9). What would you see? You would see yourself from behind, as if your own body were your character in a third-person perspective video game! After all, as far as your eyes are concerned, your body would be a remote appendage out in front, similar to your nose. After a while, you would start perceiving your visual body as *being* you again, you'd just have a new fictional vantage point (*more* fictional than the one between your eyes)—one behind your own body. That you can perceive a body to be you and yet not have a vantage point resting on that body may seem strange at first, but exactly the same thing happens when you play a third-person perspective video game.

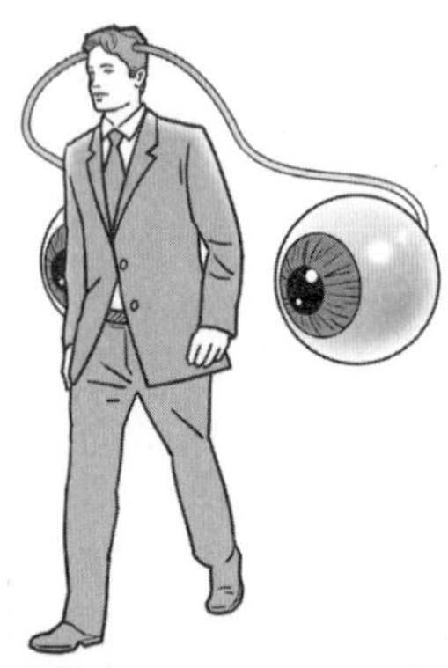

a. What your left eye sees

b. What your right eye sees

c. What you see

FIGURE 9. *(**top**) Imagine having eye stalks so that your eyes could be suspended behind you and on either side of you. (**bottom**) Or instead, imagine rigging cameras to your back and feeding the images to your eyes with goggles. (**a**) and (**b**) are what your left and right eyes would see, respectively: Your left eye sees your body to the right of its visual field, and your right eye sees your body to the left. (**c**) An illustration of what is perceived with single, unified perception. It consists of two views of yourself, each perceived to be transparent. Your vantage point appears to be situated well behind your body.*

There would be several advantages to having a pair of cameras rigged behind you that you don't experience while playing a video game. First, as we've discussed, you'd not only see your own body, but through it to the scene beyond. That is, none of the scene would be occluded by your body. That doesn't happen in video games. Second, you'll actually see two different views of your own transparent body—one from the left, and one from the right—providing better visual feedback about your interactions with the world. And third, when you play most third-person perspective video games, the camera floats behind the character in steady-cam fashion, with the camera buffered from all the moment-to-moment jerks and jumps of the character. Although one might consider this an advantage, I doubt it. In real life, your visual system is designed to interpret the optic-flow patterns that occur during all these movements. You know your body is moving not because you see your body moving, but because you see the optic patterns in the world moving. And as we will discuss in the next chapter, your visual system knows how to perceive the present when optic flow patterns behave the way they should.

Could viewing the world from behind our bodies be better than seeing the world as we actually do? Even if it *were* better, there are very good reasons why it would not be advantageous to have eyes floating perilously behind our heads. But my guess is that, even if we could somehow gain this view of the world without the long eye-stalks, it would come with a trade-off: there would be some things we could do better, but also some things we couldn't do as well. Such a third-person viewing system would mean that we couldn't see our nose, mouth, or hands as well, and we'd be less competent at coordinating their interactions with the world. However, we would be able to see our entire bodies, something we currently can't do, and this might allow us to coordinate our whole-body movements better. (Perhaps you'd finally be able to do that cartwheel you never could manage as a kid.) It could be particularly useful for those with injuries or diseases that impair coordination, for just as having visuomotor feedback of your arms can help you more quickly rehabilitate arm movements after injury, having visuomotor feedback of your whole body might allow quicker rehabilitation of overall coordination.

Whether or not you're interested in paying $1,999.95 for my third-person self-viewer, the point is that seeing a unified view of our appendages and the scene beyond them may be the reason we have eyes on either side of the head, and the accompanying binocular region, in the first place. None of this, however, gets at another key question about animal eyes: Although nearly all animals' eyes are on either side of their midline, why do some animals have eyes that face sideways, giving them panoramic vision but only a small binocular region, whereas other animals (like us) have eyes that face forward, allowing us to see only in front but giving us a large binocular region? To hope to understand this we need to get a handle on what is so special about the binocular region beyond the ability it gives us to see through ourselves, which only requires a binocular region large enough to fit our appendages. In a sense the question can be reduced to: Why are our binocular regions so much bigger than our noses? Alternatively, what about the binocular region is so useful for viewing the world beyond ourselves? Given the X-ray power we've discussed, being able to see through ourselves, it may occur to you that the same X-ray power could also extend to the rest of the world. Might we have X-ray powers for seeing through other objects, not just our own noses? And if so, could *this* be key to understanding why our eyes face forward?

Video Games for Cyclopses

Few of us have much experience with what it's like to be a cyclops. We can close an eye any time we wish, but we usually don't—and certainly not for long periods of time, much less while engaging in complex acrobatic tasks. But if you've played a first-person-perspective video game, then you *do* have significant experience with what it's like to be a cyclops, because no matter how many eyes you have, when you play such video games, you're playing as a cyclops (Figure 10). This is because video games present just a single image—the world as if seen from only one eye. Cyclopses would therefore probably be better at video games than we are, because they're accustomed to being cyclopses—they wouldn't have to adjust!

In light of the cyclopean nature of video games, playing these games can provide hints as to what is so useful about binocular vi-

FIGURE 10. *Cyclopses are just as good as we are, or better, at video games.*

sion (besides the ability to see through ourselves). Namely, we can ask ourselves what we *miss* when playing video games, and use this to help us pin down binocular vision's benefits.

Since people typically think binocular vision is about depth perception (called stereopsis), we might wonder whether it is difficult to perceive distances in these one-eyed views of scenes. And the answer is no, not really. These images are so richly realistic that they possess abundant cues as to the distances between different parts of the scene. And that's just for *single* screenshots. When you're actually playing these games, moving around, the distances are even more apparent due to a strong depth cue called motion parallax, in which objects that are nearby move more in your visual field than objects that are far away.

Our ability to expertly and acrobatically navigate realistic virtual worlds as a cyclops suggests that stereopsis may not be quite as important as it is often thought. Its importance in real life has, in fact, been surprisingly difficult to establish. People who have lost an eye lack not only a binocular field but a significant portion of their visual field altogether, yet even these people describe themselves as fully visually capable. People with only one functional eye have been pilots, race car drivers, surgeons, and pirates. It has even been difficult for researchers to find evidence that these people have more automobile accidents.

Depth perception, then, isn't much missed for those without binocular vision. Is there anything that we *do* lose when we don't have binocular vision? We have already discussed in detail one thing we lose: the ability to see through ourselves. But do we lose anything else?

I think so. While playing *Call of Duty 2*, I preferred not to run around like mad and shoot at everyone in sight without regard to my own life, but instead to actually be stealthy, and try to survive a twenty-five-minute round while dying as few times as possible, and preferably not at all. My weapon of choice was usually a sniper rifle. I would find a safe spot to lie down and patiently wait until a distant target came along that I could dispatch from my hiding place. Obvious places for a sniper to lie in wait were bushes or other leafy brush.

And that's where this cyclops ran into trouble. I noticed that it was practically impossible to see adequately beyond a bush. I could only locate targets through gaps in the leaves, so in order to see much around me, I had to constantly shift a little to the left and to the right, and the movement would give my position away. To help see how qualitatively different it feels to be a cyclops behind a bush rather than a normal human being with two eyes, look at the simulated grass shown in Figure 11. The illustration in Figure 11 shows what a cyclops sees when immersed within a tuft of grass: not much. It's not only while playing video games that people view the world as a cyclops. Real snipers, although equipped with forward-facing eyes, become cyclopses when peering through a sniper scope, and suffer a problem similar to the one I experienced playing *Call of Duty 2*—although with more serious potential consequences.

You may think this is not surprising. When you're in a bush, *of course* you can't see out except through the gaps between leaves, so how could we be better off in a real bush? Having two eyes looking

FIGURE 11. *A tiny tuft of simulated grass. If we, with our two eyes, were to actually lie behind that grass, we would be able to see through it quite well. When playing the game, however, we are effectively cyclopses and cannot see through it.*

in the same direction, however, changes things drastically—if the separation between the eyes is greater than the sizes of the leaves. You get the advantage of two views out of the bush, not just one, and this can sometimes be the difference between perceiving everything outside of the bush, and perceiving only half of it (but perhaps seeing nothing well enough to identify it).

To see what I mean, hold a pen vertically in front of your face and look at something far beyond it. If you close one eye, and then the other, you'll see that in each case the pen blocks part of your view, just as was the case with your nose. But with both eyes open, you can see everything behind the pen (as well as the pen itself). Most of us noticed this back when we were kids. Now spread the fingers of both hands and hold them up. Note how much of the world you can see beyond them when both eyes are open, compared to when only one eye is. You miss out on a lot with only one eye open, but you can see nearly everything when both eyes are.

Figure 12 shows two views from within the same tuft of grass we saw in Figure 11. Neither image shows what's beyond the grass on its own, but by binocularly fusing the two images together into one binocular image, the view improves. To fuse the images, first stare at the pair of images, then focus your eyes *through* the page (i.e., as if looking at something far on the other side of the page even though you cannot, of course, see through the page). When you do this, you should perceive two copies of each of the images in Figure 12, for a total of four images. With these four perceptual images, your goal is to get the center pair to perceptually overlap, so that there are only three images. You can do this by changing your focus gradually: focus too far beyond the page, and you'll perceive there to be four images. Focus too close to the page, and you'll perceive just two. Somewhere in between you'll see three images, and importantly, the center image will be a fusion of the ones on the left and right. Not only will the scene appear to be more three-dimensional, but you'll get an "X-ray view" through the grass, giving you a much better visual impression of what is on the other side.

Figure 13a shows the left- and right-eye views for another image, in this case a face blocked by leaves. Figures 13b and 13c illustrate the way you actually perceive the images, depending where you focus your gaze. When you focus on the face, you see the whole face and two transparent images of the leaves. When you focus on the leaves, you see the leaves with two transparent, overlapping images of the face behind them.

As we anticipated earlier, we do have a kind of X-ray power that lets us see through objects beyond ourselves. And with all due respect to Superman, our version has some advantages over his. First, no one is quite sure how Superman's X-ray vision works. It could involve some sort of dangerous radiation, like, well, X-rays. If so, the general populace is not going to be happy with you exercising that power very of-

FIGURE 12. *Left and right eye views through part of the grass from Figure 11. Not much is visible in either of these images, but if you were lying there, you would be able to see normally by fusing these two images into a unified perception. You can simulate this experience by staring through the page until the left image overlaps the right image. When that happens, you'll perceive a unified binocular view, and be able to see through the clutter.*

ten, which takes some of the fun out of having it. Second, even if you are not bothered by irradiating the general populace, you still don't want to hurt yourself, so you wouldn't want to use Superman's X-ray vision to see through your own appendages. Third, whereas Superman's power cannot see through lead, our X-ray vision is not impaired by any material. So long as an object is not as wide as the separation between our eyes—like a pen or most leaves—we can see through it (we'll discuss the conditions required for our X-ray power to work in more detail later). Fourth, our X-ray vision can not only see through clutter, but also simultaneously see the clutter itself, whereas Superman's X-ray vision—if it is similar to X-ray imaging in medicine—loses sight of much of what it's seeing through.

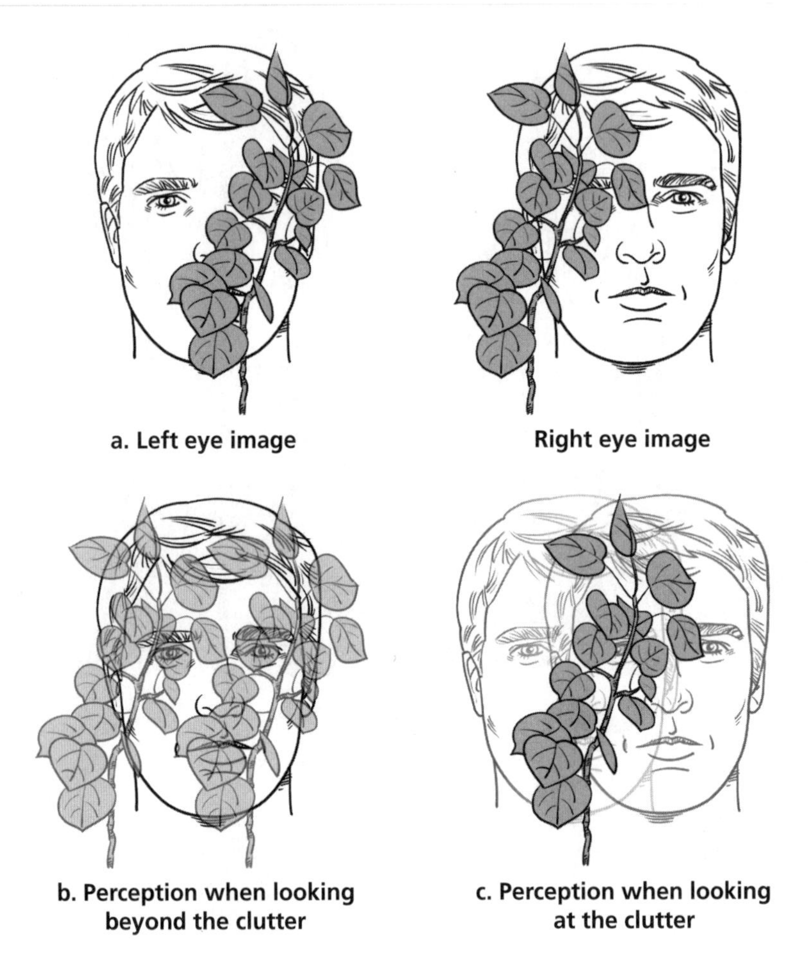

FIGURE 13. (**a**) *What the left and right eye see when looking at a face on the other side of a tree branch.* (**b**) *What you see if you focus on the face: two images of the tree branch, each transparent, and one opaque face behind it.* (**c**) *What you see if you focus on the branch: one opaque branch with two transparent, overlapping faces behind it. Whether you are focusing as in (b) or in (c), you are able to perceive the entire face and the entire leafy branch.*

Given everything we discussed earlier about seeing through ourselves, the conclusion that we have the ability to see through other objects should come as no surprise. As far as our eyes are concerned, objects in front of our eyes are just that, objects in front of our eyes, whether those objects are our own appendages, something being held by our appendages, or objects beyond our appendages. Is this ability to see through things just an accidental, unimportant side effect of the depth perception binocular vision gives us? Or could it be that my difficulty seeing through leafy clutter in video games, and my lack of difficulty seeing depth in the same games, provides a hint to the true central function of binocular vision? Could the binocular vision and forward-facing eyes of Kipling's Leopard and Ethiopian in the quote that began the chapter be for seeing Zebra and Giraffe more easily in the forest? Could all the attention to depth perception over the history of binocular vision have been beating around the bush?

Severing the World in Two

When I was a kid, I would often stare at the flower-pattern wallpaper in the bathroom, allowing my gaze to go beyond the wall. As I did so, the solid wall appeared to acquire a vertical crack down the middle, as if the single wall was severing in two. If I tilted my head, the two sides appeared to separate, one sliding slightly toward me. I was not prone to fancy, and this was a couple of decades before movies like *The Matrix*, but I did once check to see if I could walk through the small gap between the severed walls. Where on Earth I expected the gap to take me I can't recall, but of course there was no path. With too much time on my hands, I had just begun over-thinking the useful fictions generated by my brain.

You can try this yourself. Cup your hands and hold them up fairly close to your face (or find good bathroom wallpaper), and focus your eyes "through" your hands. As was the case for the pen discussed earlier, you will see two copies of your hands. There is a key difference here, however. For the pen the two images of it in your perception did not overlap, and so you perceived each to be transparent, allowing you to see the scene beyond it. But when you create a double image of your cupped hands, the two images overlap in your perception and you are *not* able to see through your hands—how could you, given that your cupped hands are blocking each eye's view? Yet your perception of each image of your hands is transparent. This may at first seem strange, because why, then, can you not see through them? But what their transparency does is allow you to see each eye's image despite the fact that they overlap each other in your perception. That is, the transparency now is useful not for seeing through to the scene beyond, but for seeing through one eye's image of your hands to the other eye's image of your hands.

The severed wall I imagined walking through into another world was, then, just another case of seeing two views of a nearby object when looking beyond it. Unless you are bored in your parents' wallpapered bathroom, however, you don't typically spend much time trying to stare through large opaque objects. So this kind of world-splitting is not something we do much, and something there is certainly little good reason for doing. However, there are natural scenarios where severing the world in two *is* useful. Hold up your finger vertically again, but *now* focus your eyes on your finger rather than on the scene behind it. You'll now see only one image of your finger, and it will be opaque, no longer transparent. But something else about your perception has also changed. While there are no longer two images of your finger, there are now two images of the scene beyond it, the left-eye image and the right-eye image. You are still not missing any of the scene—not because your finger is perceived to be transparent, but because the world beyond it has split in two (see Figure 13c). How, though, are we able to see two images of the scene, split apart inside our perceptual screen, when that screen should only be big enough for one? The brain makes this possible by letting the two images partially overlap, while rendering them each transparent so that neither occludes the other. That is, the brain uses transparency again in a useful fiction, but now that transparency is not so that we can see through the scene to whatever's beyond it, but so that we can see both images of the scene simultaneously. This is the same trick my brain was playing when I peered through the wallpaper, except that here we are peering at something closer than the wallpaper (e.g., peering at a finger, which the wallpaper is behind).

Transparency is a handy tool, not only for letting us see genuinely transparent objects, but also for depicting useful fictions of (1) our appendages, (2) objects we are seeing through to the scene beyond, and (3) even the world itself, when we are focusing on a nearby object. And note that, unlike the perception of transparency that occurs when you are actually looking at something transparent, here you can choose to focus both eyes on a target—like your pen—and make the target opaque, allowing you to see it with greater clarity. Transparency wouldn't be needed as part of the perception-construction toolkit, however, if each eye saw all the same stuff. Because one eye sees something that the other eye doesn't, the brain must figure out how to somehow render both, and uses transparency to help solve the dilemma. We turn next to this issue of unmatched images.

Unmatched Vision

Peeking around corners is not one of the things I'm most proud of, but I'm very good at it. And so are you. You may think that peeking is not the kind of thing you *can* be good or bad at. Sure, you can use peeking for good or evil, but the ability to peek itself isn't very impressive. Peeping Tom may be evil, but he's not a supervillain! Peeking, however, does require specific mechanisms in the brain—mechanisms that had to be selected for during evolution. Without these mechanisms you wouldn't be able to peek, or at least not as well. To understand why, lift up a coffee mug or magazine and peek around it with one eye. Importantly, let your non-peeking eye remain open, even though it can't see anything but your coffee mug. What do you notice? You perceive the whole scene that you're peeking at. But you also perceive your coffee mug; you perceive it to be transparent, through which you are viewing the scene beyond.

This sounds unremarkable at first, especially given everything we've discussed thus far about seeing through objects. But notice that each eye perceives something *entirely* different, yet your brain can still unify the separate images into a single perception. Throughout most of its history, the study of binocular vision has concentrated on how the brain deals with two images that only differ in the angle from which they show a scene (and out of which three-dimensional perceptions can be built). What are we to make of successful perception when the two images have no parts that correspond to one another at all? When our eyes are presented with two utterly different images, the brain cannot typically offer a unified perception. For example, if you present a checkerboard image to one eye and a spiral image to the other, the brain is incapable of building a unified perception—because there isn't one. Instead, the brain tries to have things both ways, eliciting a unified perception of the checkerboard, then the spiral, and then the checkerboard again, and so on. This is called binocular rivalry, and does seem like a good solution to a bad situation.

Why, though, do we not experience rivalry when peeking? Your brain somehow knows that *that* pair of utterly different images—a scene image in one eye and a coffee mug image in the other—is sensible, and knows just the kind of useful-fiction unified perception to build. That is, the brain knows how to deal with peeking, and more generally, it knows how to handle cluttered situations in which there is a leaf or some other obstruction in front of one eye but not the other. The lack of rivalry when peeking, when rivalry does occur in the checkerboard/spiral case, suggests that the brain is good at building perceptions when within clutter, and that it is not simply reacting in some automatic always-the-same fashion when the inputs to the right and left eye are utterly different.

And it is not just that the brain knows not to generate the perception of binocular rivalry when peeking. The brain also knows to make the coffee mug appear nearby and transparent, and the scene appear farther away and opaque. How does the brain know to do this rather than generate a perception of the coffee mug through a transparent scene instead? There is a key difference in the two images that makes this decision easy: because you are looking at the scene far beyond your coffee mug, the mug is out-of-focus and accordingly blurry; the scene is not as blurry. That's how the brain knows to see through the coffee mug to the scene, and not vice versa. In fact, Derek H. Arnold, Philip M. Grove, and Thomas S. A. Wallis in 2007 noted that the brain is very good at doing this. They pointed out that if you are peering at an object through a leafy forest, then as you move (or as the leaves blow), which eye sees the object, and which eye has a leaf in the way, may switch back and forth. But throughout, you still perceive the object, and any leaves blocking one eye's view will still be perceived as transparent. That is, which objects need to be transparent can be changed on the fly; although new leaves are constantly getting in the way, and which eye it is that is blocked is constantly changing, the brain is able to track that the blurry objects must be the ones that need to be seen through. This is not something we would expect to evolve in a creature living in a non-cluttered environment, suggesting that we have been designed with clutter in mind.

Hints as to how good our eyes are at building a single unified perception when images in the left and right eye do not match cropped up in the late 1980s when vision scientists Shinsuke Shimojo and Ken Nakayama began carefully investigating how the brain decides on the depth of an object seen by only one eye. By virtue of their work, and that of many others since, we have come to realize that the brain has an incredible understanding of the principles governing how objects occlude other objects in the world—knowledge the brain uses to build a unified perception despite receiving unmatched images. For example, when the left eye sees a marble to the left of a coffee mug, but the right eye cannot see the marble at all, it infers that the marble must be behind and to the left of the mug, and builds a perception of that.

Although the work by Shinsuke Shimojo and Ken Nakayama helped initiate greater study of how the brain handles unmatched images, a full appreciation of our ability to see through clutter has taken some time. Arnold and colleagues' work, which we discussed above, is one example; it gets at how dynamically and how well we handle situations where clutter jumps from one eye to the other. Another paper, by New York University researchers Jason Forte, Jonathan W. Peirce, and Peter Lennie, from 2002, has driven home how well we manage unmatched vision situations more complex than peeking. They presented cases where both eyes see the same clutter, but different images beyond it. They had observers look through a fence at targets, where the images seen by the left and right eyes were akin to those shown in Figure 14a. The left eye sees stripes of a scene (in Figure 14, another person's face), and the stripe-shaped gaps not seen by that eye are seen by the other eye. Figure 14b shows just the stripes seen by each eye, without showing the fence, to help illustrate exactly what view is available to each eye. The two eyes' views of the face are completely different, and the brain must take these two incomplete images and figure out how to glue them together into a unified perception of a face. As these researchers showed, and as is illustrated in Figure 14c, the brain can indeed do this. If you can fuse the two images in Figure 14a by staring through the page, then you'll perceive something roughly akin to the illustration in Figure 14c.

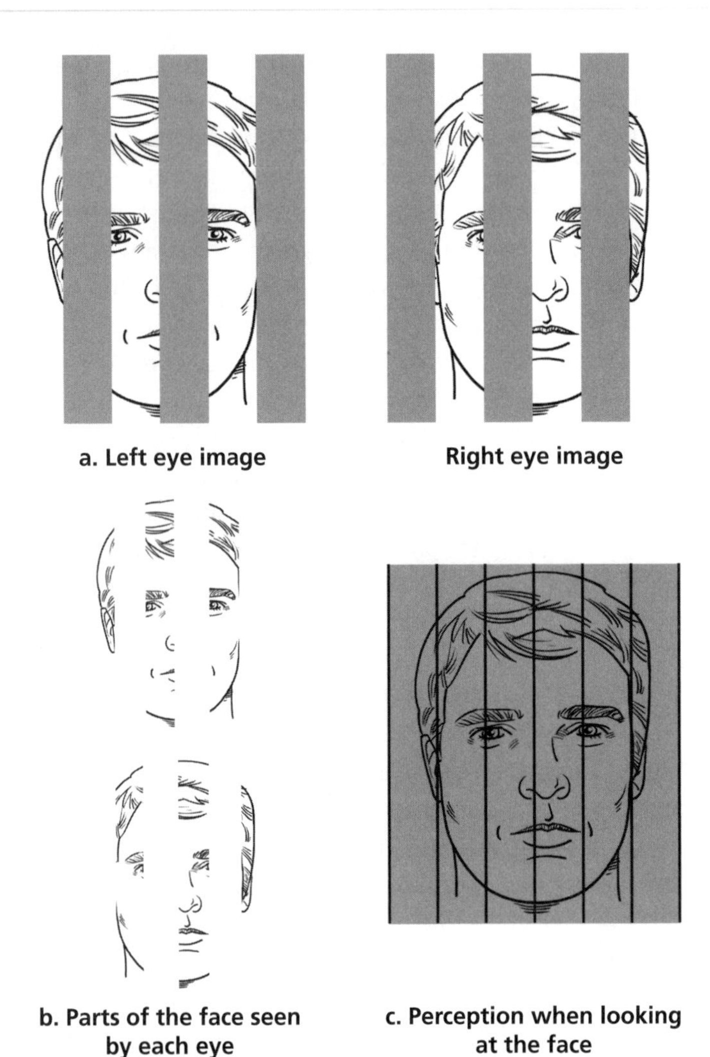

FIGURE 14. *(a) What the left and right eyes see when looking at a face through a fence. Specifically, here it is rigged so that the left and right eyes see completely different regions of the face, but in total, all regions of the face are seen. (b) The regions of the face seen by each eye, placed above one another to help emphasize the "jigsaw" nature of what the brain must integrate and unify. (c) What you see when focusing on the face. You perceive two transparent copies of the fence, through which you see the entire face. This is a special case where the left and right eyes see entirely different images, but together don't miss anything. Jason Forte, Jonathan W. Peirce, and Peter Lennie carefully investigated such jigsaw binocular perceptions at NYU and published their findings in 2002.*

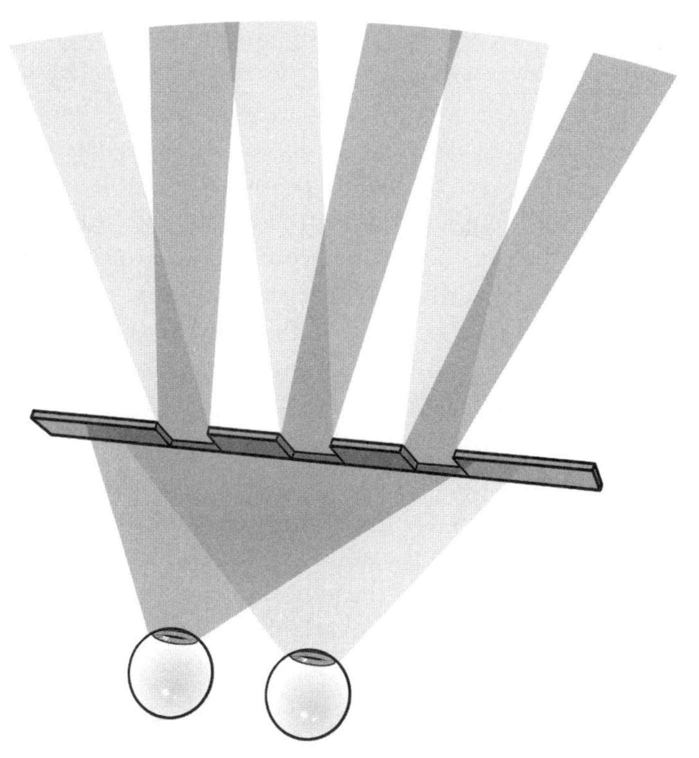

FIGURE 15. *Each eye tends to see different parts of the scene behind clutter.*

One might complain that the kind of case in Figure 14 is a bit unrepresentative because fences are a modern phenomenon. And more generally, how often do our eyes really have such different views beyond clutter? Because we now live outside of leafy forested environments, it is easy to underestimate how common this is in forests. Figure 15 shows what each eye sees beyond a layer of clutter, and we can see that each eye is seeing different parts of the scene beyond. Unmatched images like those in Figures 13a and 14a are a reality in forests, and dealing with similarly cluttered situations like these is a problem the brain had to overcome. But it was also an opportunity, because once the brain had mechanisms able to handle unified binocular perceptions in clutter, it could secure itself a superior view of the (cluttered) world. This led to the evolution of forward-facing eyes in order to expand the size of this superior view. And this is the topic we take up in detail in the next section.

But before moving on, let's anticipate the next development by making an observation about the perception illustrated in Figure 14c. In addition to seeing the entire face through the fence, our brain also perceives the fence itself, using the same two-transparent-copies trick we've seen before. Interestingly though, in this case the two transparent copies completely cover the visual field, leaving no gaps. We've seen this before too—when you peeked around your coffee mug—but Figure 14c differs in that the transparent "screen" over your unified perception is built from the views from each eye, rather than just one. In either case, these transparent screens show that our vision is not limited to seeing only one (opaque) object in any given direction. Unlike animals with sideways-facing eyes, because of our binocular field we can see up to *two* opaque objects in any given direction! Animals with forward-facing eyes don't lose the ability to see anything; rather, they gain the ability to see two *layers* of things in a cluttered visual field. It is as if the panoramic vision of sideways-eyed animals is pulled forward and overlapped to allow layered perception, i.e., a unified perception that may only take in one hemisphere of visual space (what is in front), but actually "sees" twice that because of its ability to see two layers. What makes two eyes facing the same direction so powerful when looking through clutter is this exact ability. That is, our unmatched vision is the key to unmatched vision in clutter, and that is the topic to which we turn next.

Eye Rock

Breaking my living room window is a one-rock job...if, that is, you dare to walk right up to my window. Bringing a second rock along for the job might be a sign of your professionalism, but probably won't be needed. Your chances of success are nearly guaranteed with a single throw; having the other rock won't enhance your chances. However, what if I were to leave my dog out front, forcing you to throw your rock from across the street instead? Then, having a second throwing rock would be very handy, and nearly double your chances of success.

Unless, rather than wildly chucking each rock by hand toward my window, suppose you bring along with you a high-precision rock-chucking mortar, so well-tuned that you can hit the same exact spot with two consecutive firings. Hiding behind a flamingo lawn ornament across the street, you aim the mortar and launch a rock. From your concealed position, you can't tell if the window broke. Should you fire another? Probably you should, since the rock-chucker's handbook recommends persistence, so you launch off a second rock. How much did firing the second rock enhance your chances of breaking my window, though? If you didn't change the aim of the mortar between firings, then you didn't enhance your chances at all. If the window broke on the first throw, then the second rock would not have been necessary. And if you missed the window on the first throw, then the second rock is guaranteed to also miss. So throwing two rocks in this fashion—via a highly accurate mortar—is no better than just throwing one. In fact, although two rocks were thrown, in a sense only one rock-throwing event type occurred...it just happened to occur twice. And the extra time and noise it took to fire off the second rock may have contributed to my neighbors noticing the strange person creeping around their lawn ornaments and alerting the police.

Even though you had no feedback about your first rock launch and therefore no guidance on how to better re-aim the mortar, even a *random* re-aiming would have been much better than not re-aiming the mortar at all. A random re-aiming of the mortar would mean that your two rock throws were independent of one another, i.e., the chances of success for the second throw would not have depended on the chances for the first throw. The two throws would then really have been two separate throws, rather than just one throw made twice. Re-aiming the mortar will nearly *double* the probability that you break the window with at least one of your two throws. Of course, it would be even better if you could gauge the success of your first throw before re-aiming, but my point is that even in the absence of on-the-job learning, randomness alone doubles your chances of success.

When, then, is it fine to show up for work with only one rock? When either the window is easy to hit *or* when the rock throws are not independent, or both. It is a good idea to throw *two* rocks at my window rather than just one when hitting the window is difficult *and* the second throw is a genuinely new attempt to break the window.

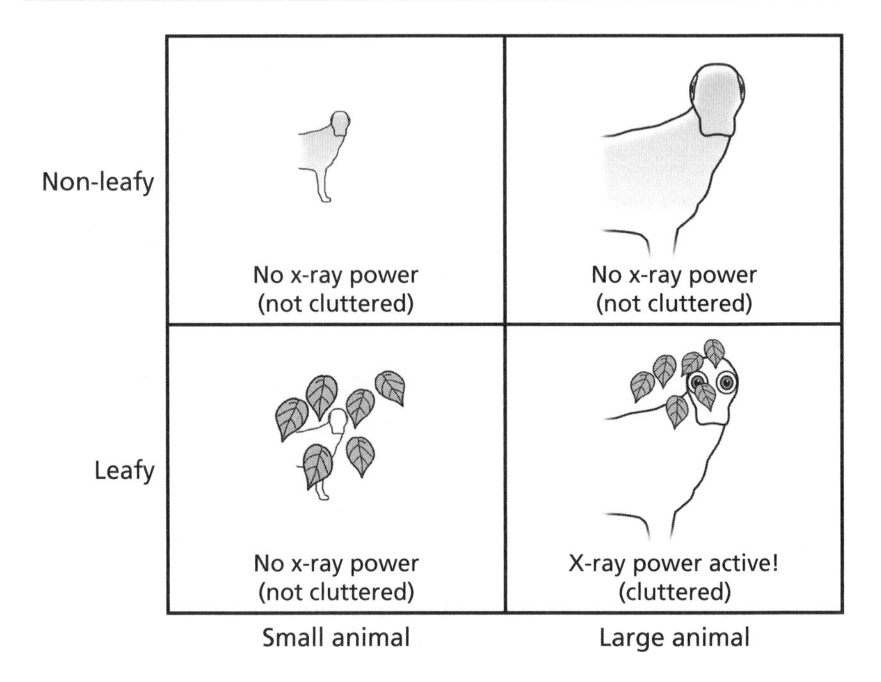

FIGURE 16. *The X-ray power of two eyes only occurs when the habitat is leafy (and thus somewhat difficult to see targets through) and the animal is large (so that the separation between its eyes is greater than the typical leaf size). If X-ray power drives the evolution of forward-facing eyes as I hypothesize, then only animals that fall in the bottom right quadrant of this table should have forward-facing eyes. We'll test this prediction later in the "Large and Leafy-Loving" section.*

Animals surveying their world are like rock-chuckers breaking windows. But rather than trying to break windows, animals are trying to successfully survey locations out in the world. Just as a window is broken so long as *at least one* rock hits it, a location will be successfully surveyed so long as *at least one* eye can survey it. And instead of the difficulty of hitting a window on any given throw, what matters now is the difficulty of surveying a location for any given eye. What makes it difficult to survey locations or regions of the world is the fact that most objects are opaque, and some locations may be occluded by those objects.

(By the way, I say *surveying* a location rather than *seeing* a location because when you gaze at an empty location [e.g., a particular spot in space three meters in front of you, and a meter above the ground], you are not really *seeing* the location but *surveying* it, in the sense that you'd see anything [any object] that moved into that spot.)

Given the close affinity of surveying and rock-chucking, it follows that one eye is nearly as good as two for visually surveying some location in the world if *either* it is easy to visually survey that location, *or* the two eyes don't have genuinely different views toward that location, or both. Concerning the first part of this "or": When *is* it easy for an animal to visually survey locations out in the world? The answer: When there is nothing occluding the animal's vision. Evolutionarily that typically meant there were no leaves or other plant material; I'll just call this type of habitat "non-leafy" for short. That is, when an animal's habitat is *not* leafy, then two eyes are little better than one for visually surveying any given location. Now to the second part of the "or": When *does* a second eye not have a genuinely different view of a location? The answer: When the separation between an animal's eyes is smaller than the size of the typical leaf (or other plant material) in that animal's habitat. Consider what happens when you try to look past an object that is much larger than the distance between your eyes. For example, consider your view in a typical city habitat, where all the objects—other people, trash cans, cars, trucks, buildings—are much larger than the distance between your eyes. Or consider what it's like to be a mouse in the forest, where leaves are as big as or far bigger than the average mouse. In such circumstances what one eye sees the other eye typically sees as well. So when the distance between an animal's eyes is small compared to the sizes of the leaves in a leafy habitat, then having two eyes with which to survey a location out in the world is little better than having just a lone eye with which to survey that location.

In sum, then, one eye is nearly just as good as two for surveying some location out there in the world if either an animal's habitat is non-leafy, or the separation between its eyes is small compared to the size of leaves in its habitat, or both. When either of these conditions apply, or both, I call the habitat "non-cluttered"; see the top row and the left column of Figure 16 (i.e., the three squares aside from the bottom right one). Binocular vision—i.e., the vision that results from two eyes looking in the same direction—gives an animal no special advantage when its habitat is non-cluttered, and so a view from one eye is all an animal needs for surveying any given location in the world. We will therefore expect animals to have small binocular regions (eyes that face more sideways)—and consequently panoramic vision—when habitats are non-leafy or the animals are small, or both.

Things become interesting when both parts of the above "or" fail—namely, when a habitat is highly leafy *and* the separation between an animal's eyes is larger than the width of the leaves. When this happens I call the habitat "cluttered"; see the bottom right box of Figure 16. In a cluttered habitat, two eyes are up to twice as likely as one eye to successfully survey a location out in the world, and these are the conditions in which binocular vision acquires its X-ray power to see better through clutter—to see the forest through the trees. That is, binocular vision has the X-ray power to see through stuff in the world only when the world is leafy and the animal is large.

If you are interested in keeping an eye on some spot out there in the world, then we now know when to point just one eye in that direction, and when to point two eyes in that direction. That is, we now know the conditions when the X-ray power of two eyes works when looking in some direction. But we are still not in a position to say what is so good about forward-facing eyes, because we need to better grasp the trade-off between forward-facing eyes and sideways-facing eyes, and this is what we'll take up next.

The Price of a Blind Behind

In our everyday lives we keep an eye on costs. Sure, I can splurge and get the extension cord I've always wanted right now, but I notice the cost, and instead leave a note to myself to get it for my wife at Christmas, killing two birds with one stone. Scientists are as cost conscious as anyone else, yet when it comes to theorizing about why animals might have evolved a certain feature, we can get so excited about our clever idea on the purported advantages of the feature that we forget to consider the costs. These costs, after all, don't come out of our wallet, and so don't cry out for attention.

This is especially true for forward-facing eyes. Ideas as to why we have forward-facing eyes frequently fail to take into account what we lose in the bargain: being able to see behind us. If, for example, forward-facing eyes have been selected for because of their wide field of three-dimensional stereoscopic vision as has traditionally been assumed, then we need to think about what makes that benefit worth the loss of our ability to see well behind us. Weighing the relative merits of stereo vision in front versus the ability to see behind is, however, a very tricky theoretical problem, because they are so different. Any scientist interested in explaining why forward-facing eyes are advantageous needs to come up with a way to compare the two. This is, alas, not easy; previous scientists' failures to do so is understandable. Nevertheless, it means that the old hypotheses may not (and I suggest do not) adequately explain why forward-facing eyes have evolved.

If forward-facing eyes are for seeing through clutter, however, then it is easier to make sense of why it is worth the price of a blind behind. When in clutter, an animal with forward-facing eyes can survey twice as much in front as the animal with sideways-facing eyes. The animal with sideways-facing eyes can survey what's in front of it and what's behind it as well, meaning that animals with forward-facing and sideways-facing eyes visually survey the same amount of space. But there is a difference. There is typically a threshold effect involved in recognizing objects beyond clutter, such that you can't recognize an object when you only see a certain fraction of it, but you *can* recognize it once you see just a little more than that fraction of it. In many cluttered situations, the animal with forward-facing eyes will be able to survey nearly *all* of what's beyond the clutter, whereas the animal with sideways-facing eyes will be able to survey only half. The animal with forward-facing eyes will easily recognize objects behind the clutter that the animal with sideways-facing eyes will have trouble with; the ability to survey twice as much in front will lead to many situations where the animal with forward-facing eyes will see a fraction of an object that falls above the threshold for recognition, where the animal with sideways-facing eyes will see a fraction that falls below that threshold, and fail to recognize the object entirely. And having informative vision only up front is better than having uninformative vision all the way around. This won't always be the case; sometimes the clutter will be so dense that neither kind of animal will recognize anything. But in a case like that, sideways-facing eyes would be little better than forward-facing eyes.

Another way to compare the advantages of forward-facing eyes to sideways-facing eyes is to tally the regions around the animals in which they are capable of recognizing objects. Let's imagine that there are two layers of clutter. Recall that when two eyes see well past a layer of clutter, the view behind the clutter tends to be seen by only one eye, as illustrated in Figure 15. Roughly speaking, then, the view seen by two eyes through the *second* layer of clutter is akin to what a single eye would see if it *only* had to look through one layer of clutter. If we are imagining, as we are, that the clutter is such that a single eye sees too little beyond it to recognize much of anything, then binocular vision is effectively blind past the second layer of clutter. If the second layer of clutter is situated very closely behind the first, then forward-facing eyes will have little advantage, because its ability to recognize objects beyond a layer of clutter better than sideways-facing eyes will only work for the thin region between the first and second layers of clutter. On the other hand, if there is no second layer of clutter at all, or the second layer of clutter is quite far off, then the animal with forward-facing eyes will have a nearly infinite advantage, being able to recognize objects all the way to the horizon or to the second, far-off layer of clutter beyond the first, whereas the animal with sideways-facing eyes would be confined to recognizing objects in its vicinity before the first layer of clutter.

Figure 17 shows an in-between case, a forest with a highly regular arrangement of leaves. Yes, the "forest" in the illustration is not quite what a real forest looks like, but it nevertheless will help to demonstrate a typical or canonical advantage one can expect from forward-facing eyes in cluttered habitats. The gray in the figure shows the regions where the animal would be able to recognize an object were one to suddenly appear there: the visual-recognition regions. For the animal with nearly sideways-facing eyes depicted in Figure 17a, its small binocular region allows it to see through a layer of clutter and recognize objects in that "spotlight" region in front. If the animal had entirely sideways-facing eyes—that is, if it did not have a binocular region at all—the gray visual-recognition region would fall entirely within that first hexagon, in which the animal is standing.

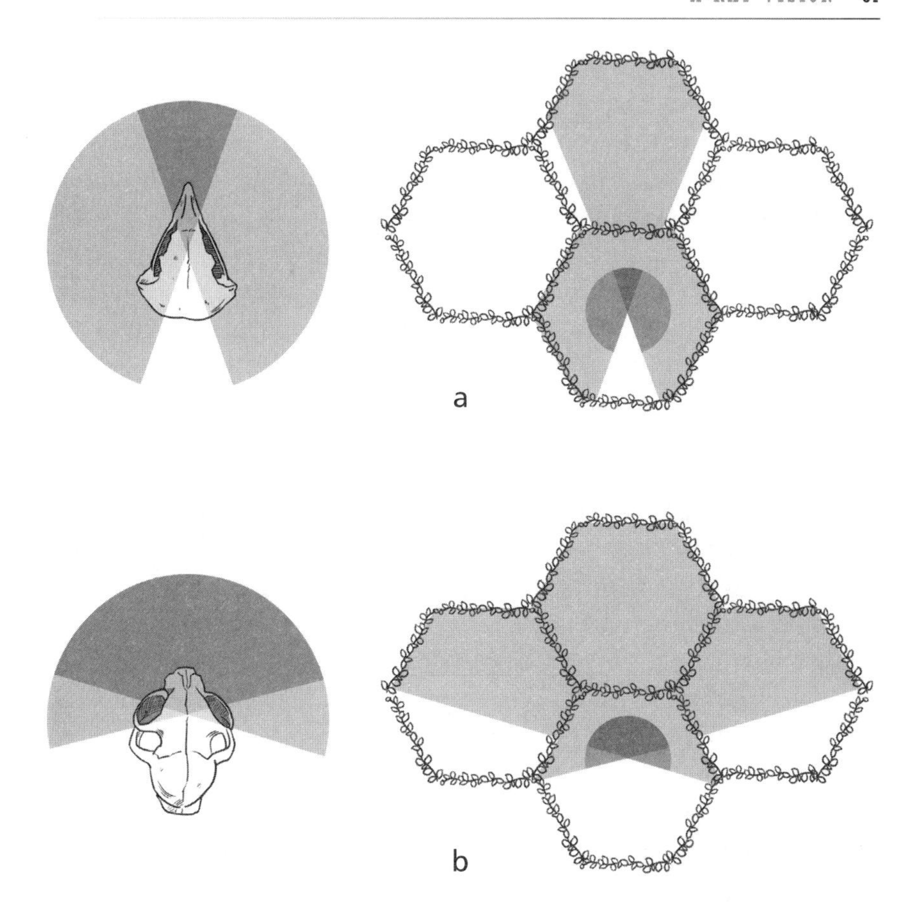

FIGURE 17. *(a) A skull of an animal with nearly sideways-facing eyes is shown on the left. Overlapping semicircles show the visual fields of each eye and the small binocular region in front. The image on the right illustrates the regions within which the animal is capable of recognizing objects were they to appear. I assume here that the clutter is thick enough that the binocular region allows it to typically recognize objects through clutter, but that a single eye's view sees too little of any object to recognize it. This animal's binocular region acts as a kind of "spotlight" shining through the clutter. (b) A skull of an animal with nearly forward-facing eyes is shown on the left. Overlapping semicircles show the visual fields of each eye here, as well. The image on the right illustrates the regions within which the animal can recognize objects. Because its binocular field is so large, the total region in which the animal can recognize objects in front is much larger than that of the sideways-facing eyed animal in (a). That is, the amount of gray in (b) is much greater than in (a), illustrating the advantage X-ray vision gives us even over the ability to see behind us in cluttered environments.*

The animal with forward-facing eyes in Figure 17b, however, has a wide binocular region, allowing it to recognize objects that fall anywhere within the three hexagons in front of it, as well as within the forward half of the hexagon in which it is standing. It can recognize objects in a total of 3.5 hexagons, whereas the animal with sideways-eyes can recognize objects in only one.

However, this actually underestimates the advantage to animals with forward-facing eyes because the little hexagonal world in Figure 17 is flat. The real world is three-dimensional and this is especially important to take into account in forests, where much of the realm of interest is often above or below. In *this* case we'd have to imagine the forest as a clump of balls rather than a flat tiling of hexagons, in which each "ball" surface is made of clutter. Because you can clump twelve balls around any single ball, animals with forward-facing eyes could recognize objects within the six balls in front of them, along with the front half of their own ball, whereas animals with sideways-facing eyes could recognize objects only within their ball. That's 6.5 times as much of the world that is visually recognizable to animals with forward-facing eyes as compared to animals with sideways-facing eyes.

You can think about the advantage of forward-facing eyes in terms of trying to start a fire with a magnifying glass on a somewhat cloudy day. After finding good kindling, you have a choice. You can rig two magnifying glasses to point at two different pieces of kindling, but you will probably just end up with two slightly warm blades of dry grass. That's what sideways-facing eyes in clutter gives you: slight views of the world around you. Or, you can aim both lenses at the same spot on one piece of kindling, using the combined energy to raise the temperature in that one spot above the threshold needed to get a fire going. Having forward-facing eyes in clutter is like focusing two magnifying lenses at the same spot: you can focus the energy of vision to burn through clutter and provide a way to see what's beyond.

Large and Leafy-Loving

Leaves, clutter, X-ray vision, magnifying glasses! A creative story, you might say, but why get into all this when everyone already knows that forward-facing eyes are for predators—for the purpose of getting a better three-dimensional stereo view of what's in front of them, all the better to eat you with. If you have any doubt about what "everyone knows," just read the Wikipedia entry for binocular vision, which states (at the moment), "Other animals, usually predatory animals, have their two eyes positioned on the front of their heads, thereby reducing field of view in favour of stereopsis."

The idea that forward-facing eyes developed to help predators goes a long way back (at least to scientist Professor Gordon Lynn Walls in 1942), and was put forth strongly in the 1970s by the anthropologist Matt Cartmill, now a professor at Duke University. There are at least two variations on this theme that have been discussed in the literature aimed at helping this theory better explain the range of eye designs found among mammals. They each hypothesize that it is hunting at *night* where forward-facing eyes are most important. One variation argues that forward-facing eyes are not for three-dimensional vision per se, but more specifically for the ability to break camouflage in the dark, just as when you look "through" a Magic Eye image and an otherwise invisible object suddenly pops out to you. The other variation (attributed to visual neuroscientists Professors John D. Pettigrew of the University of Queensland and John Allman of Caltech in the late 1970s) suggests that perhaps forward-facing eyes are useful in reducing the amount of blurring that occurs when light from the front traverses through the lens in the eye, providing a sharper image at night.

The biggest problem with the predator-prey theory is one we talked about earlier: that although it is a fair hypothesis to put forth, it does not explain why forward-facing eyes are worth the price of being blind behind. But there are other difficulties, as the anthropologist Professor Robert W. Sussman of Washington University in St. Louis pointed out in the early 1990s. The first is that there are numerous predators with largely sideways-facing eyes, including many small carnivores like mongooses and tree shrews, not to mention snakes, birds, and nearly all fish. The second is that there are many non-predators who do have forward-facing eyes. For example, primates are typically opportunistic predators at most, and many primates aren't predators at all, yet they have forward-facing eyes. And fruit bats, which eat, well, fruit, have largely forward-facing eyes. Supporters of the predator-prey hypothesis have come up with patches for these holes (such as that perhaps primates have forward-facing eyes because of a predator ancestor), but the hypothesis is still not an elegant fit to the data. By the end of this section we'll see yet another problem with the predator-prey hypothesis: it does not explain why among some groups of animals, larger animals tend to have eyes that face farther forward than small animals, regardless of whether or not they are predators. Supporters of the predator-prey hypothesis have had to come up with other hypotheses attempting to explain this (often referring to hypothesized constraints on how eyes fit into skulls), making this hypothesis even less parsimonious. Professor Sussman believes that, for primates and bats at least, forward-facing eyes co-evolved with flowering plants—a view consistent with my X-ray idea.

Is there any other support for my X-ray idea? And what exactly does my X-ray idea predict, anyhow? In short, it predicts that it is the large and leafy-loving animals that should have forward-facing eyes. Why? Because it is the large and leafy-loving animals that actually receive a benefit from X-ray vision, as we saw on the bottom right of Figure 16. And because, as I explained at the end of the previous section, an animal with functioning X-ray vision wants as much of it as possible—even at the expense of seeing behind itself—because it enables greater recognition of objects in the world around it (e.g., see Figure 17 again). Is this prediction true?

One might first note that *we* are large and, evolutionarily speaking, leafy-loving, in that we primates have tended to be selected for forest-living. We also have forward-facing eyes, fitting the prediction. So far, so good. But do *all* animals follow these guidelines? That is, are the animals with forward-facing eyes generally the large and leafy-loving ones?

Thus far, I've tended to talk about animals with sideways-facing eyes and animals with forward-facing eyes as if there were only these two kinds of animals. In reality, you find animals everywhere in between these two extremes. Scientists use the word "convergence" to describe the degree to which an animal's eyes face forward. A convergence of zero degrees means the eyes are completely faced to the side, and a convergence of 90 degrees means the eyes are completely faced forward. Figure 18, based on the dissertation work of the anthropologist Dr. Christopher Heesy, now at the Department of Anatomy at Midwestern University, shows how convergence varies with body size across a set of 319 mammals. The figure is useful for illustrating the wide range of degrees of convergence that occur in nature, from completely sideways-facing to completely forward-facing—from zero degrees convergence all the way up to ninety. *That's* what we need to explain. That is, we need to be able to explain why some of those animals are high in the plot, and some are low . . . and why some are in between. Does my X-ray vision theory help us get a handle on this?

Let's separate animals cleanly into two groups: leafy-loving animals and non-leafy-loving animals. Non-leafy-loving animals should have fairly sideways-facing eyes, and this should be true no matter their size. We already saw this in the top row of Figure 16. For leafy-loving animals, however, their eye position depends on their body size, as summarized in the bottom row of Figure 16. Small leafy-loving animals should have sideways-facing eyes because, to them, the leaves are too large for them to use X-ray vision to see through, whereas larger leafy-loving animals should have progressively more forward-facing eyes. These are the central predictions suggested by my X-ray hypothesis regarding the evolution of forward-facing eyes, and are summarized in Figure 19a.

Do the animals in Figure 18 obey this prediction? To test this, I took that data set of 319 mammals and categorized each of the mammalian orders (e.g., primates, carnivores) as "leafy," "semi-leafy," or "non-leafy" using *Grzimek's Animal Life Encyclopedia* and the *Animal Diversity Web*. All the species in an order were deemed "leafy" if the typical species in that order inhabits leafy habitats like forests. All the species in an order were deemed "non-leafy" if the typical species in that order inhabits open habitats like plains, savannas, and so on. There were some mammalian orders in between, not easily lumped into one category or the other, and these were placed into an intermediate category called "semi-leafy."

Figure 19b shows the data split into these three groups, and one can immediately see the resemblance to the prediction in Figure 19a. The species from leafy orders have convergence that increases with body mass, whereas the species from non-leafy orders have low and

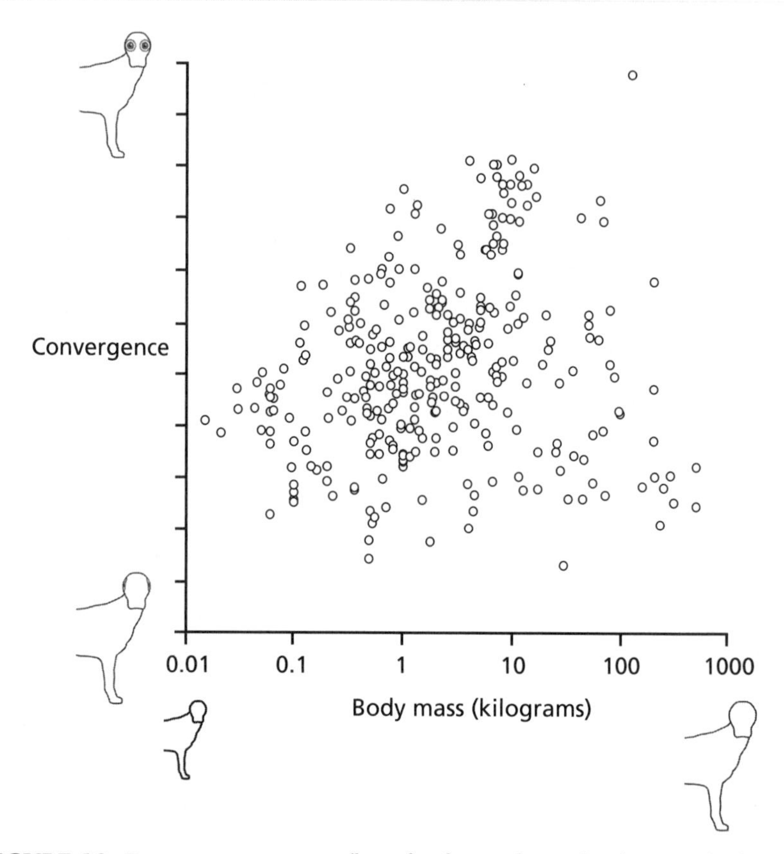

FIGURE 18. *Data on convergence (how far forward eyes face) versus body mass for 319 mammals, drawn from Dr. Christopher Heesy's dissertation work. The data are all over the map, which helps communicate the difficulty in trying to explain why some animals have forward-facing eyes and some animals have sideways-facing eyes. The only salient feature in this graph is the lack of points in the upper left, i.e., the lack of very small animals with forward-facing eyes. This is expected if forward-facing eyes evolved for X-ray vision, because when you're small, your eyes can't see around anything, and so forward-facing eyes don't allow you to have X-ray vision. Other than this region of missing data, the data here are a frightening mess. In a moment we'll see how the data nicely separate when we split the animals into those that are leafy-loving and those that are not.*

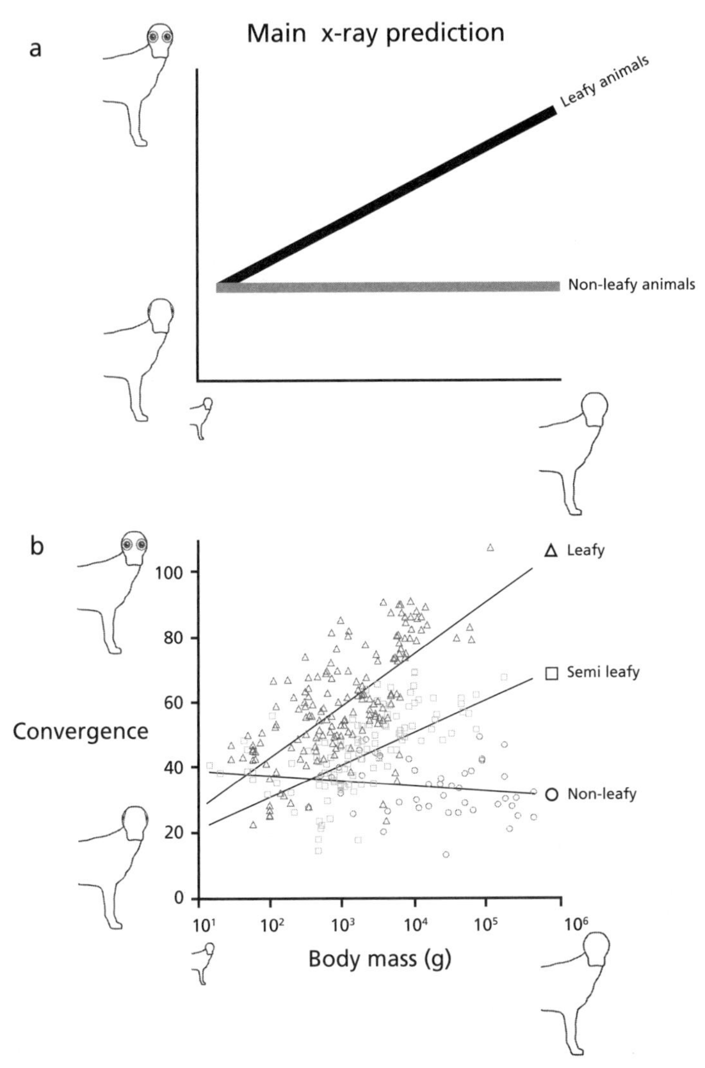

FIGURE 19. (**a**) *Predictions for leafy and non-leafy animals suggested by my X-ray hypothesis. If X-ray vision is what has driven the evolution of forward-facing eyes, we would expect small leafy animals to have fairly sideways-facing eyes and larger leafy animals to have forward-facing eyes. Non-leafy animals, however, would have fairly sideways-facing eyes regardless of their size.* (**b**) *The same data as in Figure 18, now separated into leafy, non-leafy, and semi-leafy. Species were categorized on the basis of the typical habitat for members of their order. For example, all primate species are categorized as leafy, because primates are generally leafy-loving. One can see the predicted pattern in the data. The same trends are also found when species are categorized on the basis of their own habitat.*

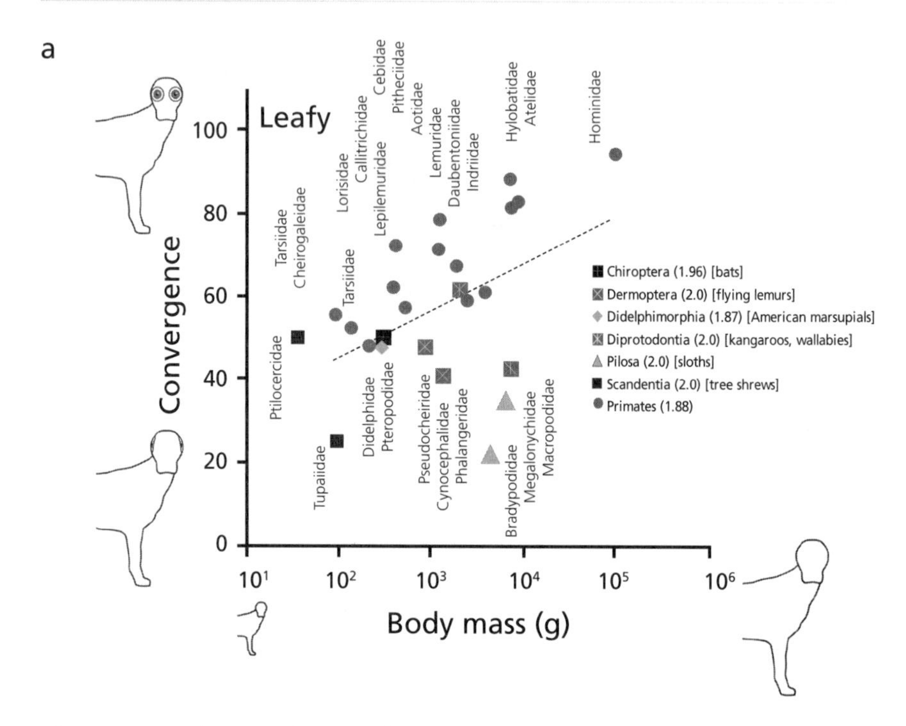

FIGURE 20. *The same data are shown as in Figure 19, but the points here are for families (groups of related species) rather than species, and spread over three plots: (**a**) leafy families, (**b**) semi-leafy families, and (**c**) non-leafy families. In each plot the shapes of the data points indicate the mammalian order the family is in.*

constant convergence. The species from semi-leafy orders are in between. In Figure 20a, 20b, and 20c I have shown the three groups of data in more detail for the interested reader. Here, each labeled data point represents a family rather than a species, and the legend specifies the mammalian order. The little numbers following the names of the orders show the average "leafiness" across the order members; each species was assigned a 0 if it prefers non-leafy habitats, a 2 if it prefers forests, and a 1 if both were mentioned in its description.

Forward-facing eyes, then, appear to be designed for X-ray vision! Whenever an animal would be benefited by X-ray vision, its eyes tend to be swiveled forward. And whenever an animal wouldn't be benefited by X-ray vision, its eyes tend to be faced sideways. *This* is the best argument that our own forward-facing eyes are primarily for

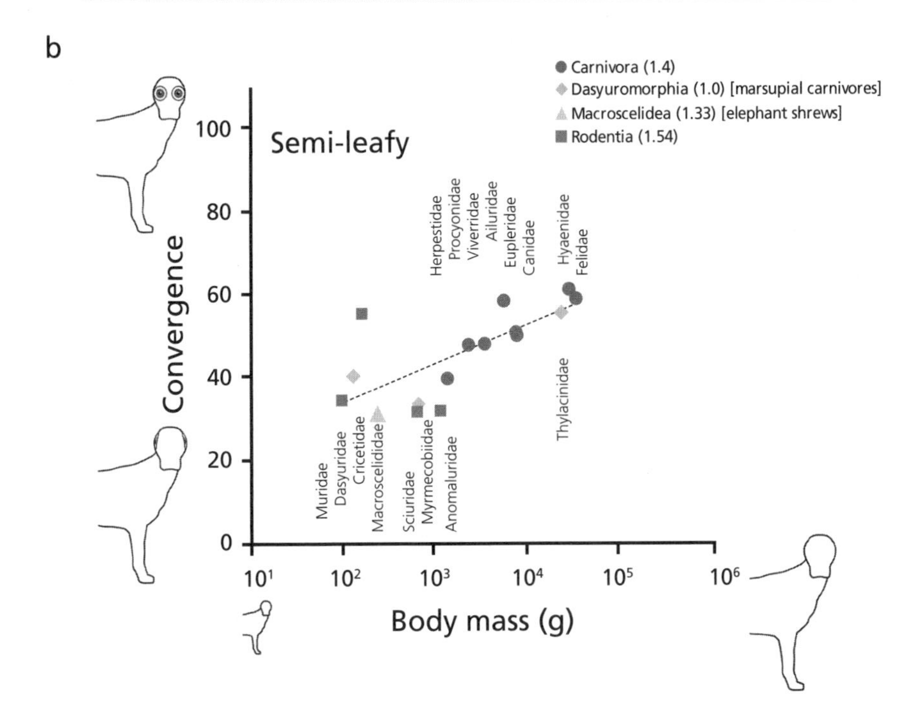
b
Carnivora (1.4)
Dasyuromorphia (1.0) [marsupial carnivores]
Macroscelidea (1.33) [elephant shrews]
Rodentia (1.54)
Semi-leafy
Convergence
Body mass (g)
Herpestidae
Procyonidae
Viverridae
Ailuridae
Eupleridae
Canidae
Hyaenidae
Felidae
Thylacinidae
Muridae
Dasyuridae
Cricetidae
Macroscelididae
Sciuridae
Myrmecobiidae
Anomaluridae

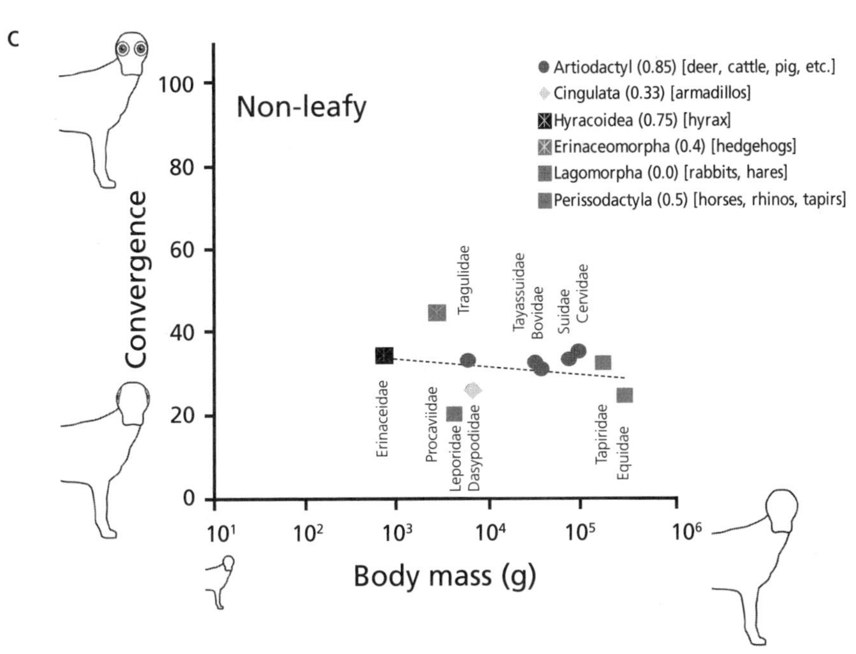
c
Artiodactyl (0.85) [deer, cattle, pig, etc.]
Cingulata (0.33) [armadillos]
Hyracoidea (0.75) [hyrax]
Erinaceomorpha (0.4) [hedgehogs]
Lagomorpha (0.0) [rabbits, hares]
Perissodactyla (0.5) [horses, rhinos, tapirs]
Non-leafy
Convergence
Body mass (g)
Tragulidae
Tayassuidae
Bovidae
Suidae
Cervidae
Erinaceidae
Procaviidae
Leporidae
Dasypodidae
Tapiridae
Equidae

the powerful X-ray vision they give us. That's why those who have lost an eye do just fine in our modern non-cluttered world, and why we can play most first-person-perspective video games effectively so long as they're set in a non-cluttered simulated world. Just don't let a cyclops be your guide on a rainforest hike! *That's* where the one-eyed (in addition to anyone whose eyes faced sideways) would have trouble.

So is there any connection between predators and forward-facing eyes at all? While it may be clutter rather than predatory behavior that explains forward-facing eyes, it's possible that some predators have a greater tendency to favor leafy habitats. It may be, then, that one finds some predators with forward-facing eyes not because they need the three-dimensional stereo vision we get from our binocular field to catch prey, but because of their leafy-loving nature.

Where does this leave the three-dimensional stereo vision we get within our binocular field? Am I arguing that it's not useful? Of course not. The question is not whether it is advantageous to have 3-D stereo vision in our binocular field. Even animals with sideways-facing eyes have a thin binocular region possessing stereo 3-D. The question is whether this kind of 3-D vision is key to understanding the evolution of forward-facing eyes—and that doesn't seem to be the case. In fact, if forward-facing eyes were designed to expand the parts of our visual field devoted to 3-D stereo vision, then we would expect the *opposite* of what is found in Figures 19 and 20. Two eyes in a cluttered environment produce very different views of the world, but to build an accurate stereo 3-D perception you need two largely similar views (from differing vantage points). Building a stereo 3-D perception from two forward-facing eyes *does*, however, work in a non-cluttered world—exactly when X-ray vision fails. X-ray vision and stereo vision are opposites. The fact that forward-facing eyes are found in animals living in clutter (i.e., the large and leafy-loving animals)—an environment in which stereo 3-D perception is the most handicapped and X-ray vision prospers—implies that X-ray vision, not stereo vision, is the reason for eyes that face forward.

Future of Forward-Seeing

At the start of this chapter you were transformed into a squirrel—small, terrified, and wishing you hadn't egged on your friend to use his "stupid toy wand." You noticed while frantically fleeing from the chimps that it seemed as if the chimps could see much better than you—you never saw a single chimpanzee, yet they clearly saw you well enough to breathe down your neck for half a mile. Your new view was handicapped in a way that you are sure it would not have been had you been climbing in the upper canopy using your old human body and eyes instead. What changed? What changed is the clutter—namely, your size in relation to the size of the leaves. As your normal self, X-ray vision would have let you see through the leaves. But as a squirrel, you had a smaller body, smaller head, and therefore smaller distance between your eyes, so X-ray vision was consequently less effective. Not only did you become bite-sized, you became significantly blinder, and the size of your binocular field was much reduced as well. Lucky for you this was just a nightmare tale, not reality. You have no magic-wand-yielding friends, and the risks of miniaturization are vanishingly small.

However, there is another real-life risk to your superpower, one that has the potential to affect us all. Although you are unlikely to shrink, what if the leaves were to *enlarge*? The question may sound a wee bit paranoid at first, but it is effectively what would happen were you to move from a forest with small leaves to one with larger leaves. It's also what would happen if you were to move from a forest habitat to any modern human environment, where leaves have been replaced by solid opaque walls, couches, cars, houses, and buildings. Although most of us city folk have never lived in the woods, evolutionarily this is the exact move our ancestors made. Rather than us shrinking, like the title character of *Alice in Wonderland*, the visual obstructions in our human habitats enlarged. Our (and Alice's) living environments have shifted from the bottom right of Figure 16 to the bottom left—from the "cluttered" category to the "not cluttered" but leafy category. We are now, effectively, cyclopses—each of our eyes sees an only slightly different view, leaving us with *half* our potential ability to see, survey, and recognize objects around us. We only need our X-ray power on the rare occasion when we are hiking in dense wood or peeping at the neighbor from the bushes.

In our modern environment, we are actually worse off than animals with sideways-facing eyes. It seems utterly humiliating: aliens observing us from the hangars of Nevada would be aghast that we have had eyeglasses for more than 700 years to improve our forward vision, and telescopes for more than 400 years to extend that vision farther, yet we have displayed no interest at all in rectifying our greatest visual failing, the fact that we are completely blind behind us. Perhaps it is *because* we cannot see behind us that we cannot fathom this as a deficiency. Unlike the need for glasses, where we can see well enough to know there is something there we should be seeing more accurately, our inability to see behind us is not something we think of as needing fixing. Since we never see anything there at all, it doesn't occur to us that there is anything in need of improvement. Rear-view mirrors on automobiles are the one exception, but for some reason, we haven't realized that a posterior view would also be useful when sitting or walking.

How might we go about correcting our vision? The most biologically natural solution would be to copy what most animals do when they are in uncluttered habitats: have our eyes face sideways, like in Figure 21a. (One can assume those smug aliens probably have sideways-facing eyes, unless, coincidentally, their world happened to have leafy brush as well.) In the very long run—e.g., after tens of millions of years of city living—perhaps we will eventually evolve sideways-facing eyes (and car accidents and muggings from behind will significantly drop). But is there anything we can do right now to make our forward-facing eyes see sideways? Probably not. Even if you wore cameras on the sides of your head that fed images to each eye, your brain wouldn't know what to do with them. Your brain only knows how to build a unified perception from the input of two eyes facing the same direction. And when the brain can't build a unified perception, it gives up, and you experience rivalry, as we discussed earlier in the chapter. Placing rear-view mirrors on each side of your head is another option; you could adjust to this, but it is hardly the kind of input the brain is designed for either.

A related alternative would be to have rear-facing cameras on each side of your head that feed into little computer display screens on your nose; the right side of your nose could have the right-behind view, and the left side of your nose could have the left-behind view. This would nicely utilize the fact that part of each of your eyes' visual field is blocked by your nose anyway, so you could get a view of what's behind you without losing anything up front.

Seeing behind ourselves is a worthy goal, and some technological progress may be possible along the lines I just mentioned, but we have to realize that our brains are not enthusiastic about the idea. They're ready and willing for a better view up front, however, so another tack to improve our vision is to give the brain what it wants and let the brain's X-ray power do its magic.

One way to do this is to respond to the enlarged "leaves" of our modern world by enlarging ourselves, especially the distance between our eyes. Imagine evolving long eye stalks, like those shown in Figure 21b. Unlike with sideways-facing eyes, we can easily use technology to give us wide eyes. We just need to place forward-facing cameras out on either side of our heads and feed the images to our two eyes via binocular goggles of some sort. This is similar to the eyes-on-stalks in Figures 8 and 9 much earlier in the chapter, but the point here is not to see through our appendages or ourselves, but to be able to use our X-ray vision power to see through more of the modern world. As I discussed then, handling these kinds of wide-eyed inputs is trivial for the brain, because as far as the brain is concerned, the inputs are coming from its regular eyes. The only consequence is that the brain would at first consider the world to be physically smaller than it actually is; a car might appear to be a toy at first (i.e., you would at first think there was a toy car only centimeters out in front of your face, rather than a real car meters out in front of you). But you'd quickly adjust to that. Figure 22 shows a simple case in which eyes separated by a distance as wide as a building would be able to see through something as large as a sailboat.

Imagine, for example, a policeman in Times Square on New Year's Eve wearing a helmet with two forward-facing cameras placed farther apart than the width of a human body. Other bodies would now play the part of leaves in a forest, placing the policeman in

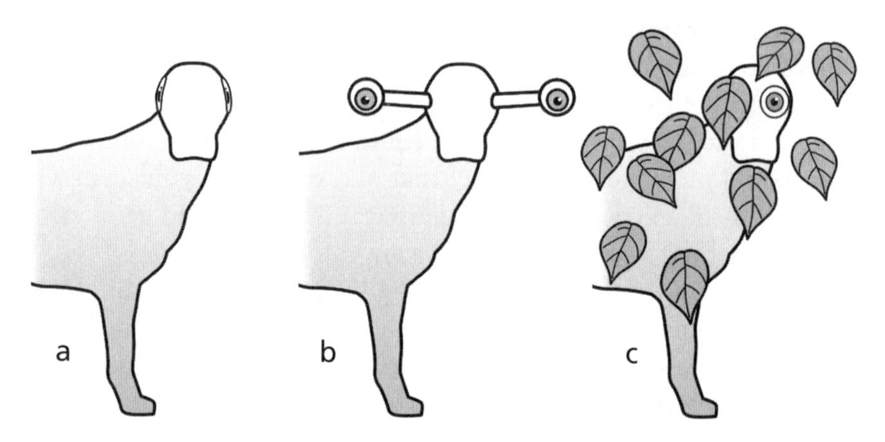

FIGURE 21. *Our eyes in the future. Our current forward-facing eye configuration at the current eye separation is not well-suited for our modern cluttered world. Shown here are three solutions.* (a) *We eventually evolve sideways-facing eyes.* (b) *Alternatively, we evolve a much wider separation between the eyes, which would require stalks of some kind.* (c) *Finally, we make our forward-facing eyes useful again by altering our environment to be highly leafy, with clutter smaller than the separation between our eyes.*

just the kind of cluttered environment his visual system is programmed for. He would see much better in the clutter-crowd than a policeman with sideways-facing eyes. And note that, although there are no policemen with sideways-facing eyes, if there *were* one, he would see twice as well as a regular policeman with regular eyes. This means that the visual payoff for the wide-eyed policeman is even greater compared to a regular policeman than the payoff is for an animal with forward-facing eyes compared to an animal with sideways-facing eyes. That is, the wide-eyed police officer in the crowd is making a visual jump equivalent to the one a cyclops makes in becoming an animal with a pair of forward-facing eyes in a cluttered environment. Compared to the civilians in the crowd, this wide-eyed policeman would see with doubly-super X-ray vision *through* the crowd.

We can also envision a new way of seeing while driving. Rather than having our eyes inside the car—akin to an animal having its eyes inside its mouth, as we discussed earlier—we can place a for-

a. What your left eye sees

What your right eye sees

b. What you see

FIGURE 22. (a) *Left- and right-eye images of a skyline with a boat in front, in which the eyes have a much wider separation than normal.* (b) *With such a large distance between the eyes it is possible to see through much larger objects than usual. In this case, we can see through an entire sailboat. The point is that by artificially magnifying the distance between our eyes, we can enhance our X-ray vision to work even in modern environments where the "leaves" are now people, cars, planes, boats, and buildings.*

ward-facing camera on either side of the car, and feed these views into the driver's eyes, as with the tractor in Figure 8. Our brains would happily take these inputs and build a single, fused perception of the world. And by doing so, we would not only be able to see the front of our vehicle better (and see through it as if it were transparent), but we would also be able to see through other cars and car-sized objects. Rather than craning our necks to see past the sport utility vehicle in front of us, we would be able to effortlessly see through it to the traffic beyond.

Finally, imagine a security officer watching for approaching threats to a building. The two forward-facing cameras here could be separated by an entire building width (or more), allowing the security officer to see a single, fused perception of what is in front of the building. With his eyes separated at such a distance, he would be able to see through any building-sized (as well as car- and people-sized) objects in front of the cameras. Rather than watch several camera videos at once, which is a very unnatural act, the security officer could look at the scene just as he would were he in the forest where his eyes evolved.

Artificially amplifying the distance between our eyes is therefore one way to deal with our modern visual dilemma. We might wonder whether evolution has tried to do this. Imagine you are a small animal—just a bit too small to get any benefit from your small region of binocular vision (i.e., X-ray powers)—living in a leafy habitat and have sideways-facing eyes. If you could evolve slightly more separated eyes, you would then be able to tap into X-ray powers using that binocular region and begin to view your environment better. Exactly how large a separation you need would depend on the distribution of leaf sizes in your specific habitat. But generally, we would expect small leafy-loving animals to have greater eye separation than similarly sized non-leafy animals. Figure 23 shows how the separation between the eyes varies with body size for leafy and non-leafy animals, and one can see that the extrapolated line showing the distance between the eyes for small leafy animals is higher than the extrapolated line for small non-leafy animals at the smallest sizes. Figure 23 also shows variability in eye separation within each kind of animal, and this variability is greater for leafy animals than it is for non-leafy animals. This variability is greater because, whereas the distance between the eyes in non-leafy animals (since they tend to have sideways-facing eyes) is tied closely to head size, the distance between the eyes in leafy animals (since they tend to have forward-facing eyes) can be more finely tuned in response to the range of leaf-sizes in an animal's niche. Therefore, the data hint that evolution may have manipulated not only the direction in which eyes point, but also the distance between the eyes, in order to make use of X-ray vision.

Also, notice that very few leafy animals have distances between their eyes wider than the width of the average leaf size (i.e., there are few data points for leafy animals that appear above the dotted line representing average leaf size), because further increases in size would not enhance their X-ray powers any further. After all, the Times Square policeman needs his eyes to be separated only slightly more than the width of a human. Having them any wider would provide no extra X-ray capabilities in the crowd and only be a severe hindrance, to say the least, in moving through the crowd ("Hey, you're stepping on my optic nerve!!"). In Figure 23, the separation between the eyes among leafy animals tends to stay within the sizes of the leaves found in forests. For non-leafy animals, on the other hand, the separation between the eyes does not appear confined to the "leaf regime" (the sizes of leaves in most environments) at all.

Thus far we've discussed two ways to improve our vision to deal with the fact that we no longer live in clutter: quickly evolve either (sideways-facing) fish eyes or eye stalks. In the case of eye stalks, at least, we can mimic the concept with technology. Technology won't help us with fish eyes, however, because our brains aren't designed to handle panoramic inputs. But there is a third way to let us take in more information about the world through our eyes. It only requires that we change the world...

...by adding clutter. Now, simply making our world cluttered would actually *reduce* our view of the world in comparison to our current view. Even if the clutter was rigged so you could always see through it, you'd see only through *one* layer; you wouldn't see in stereo 3-D beyond that layer, and you wouldn't see much at all past the

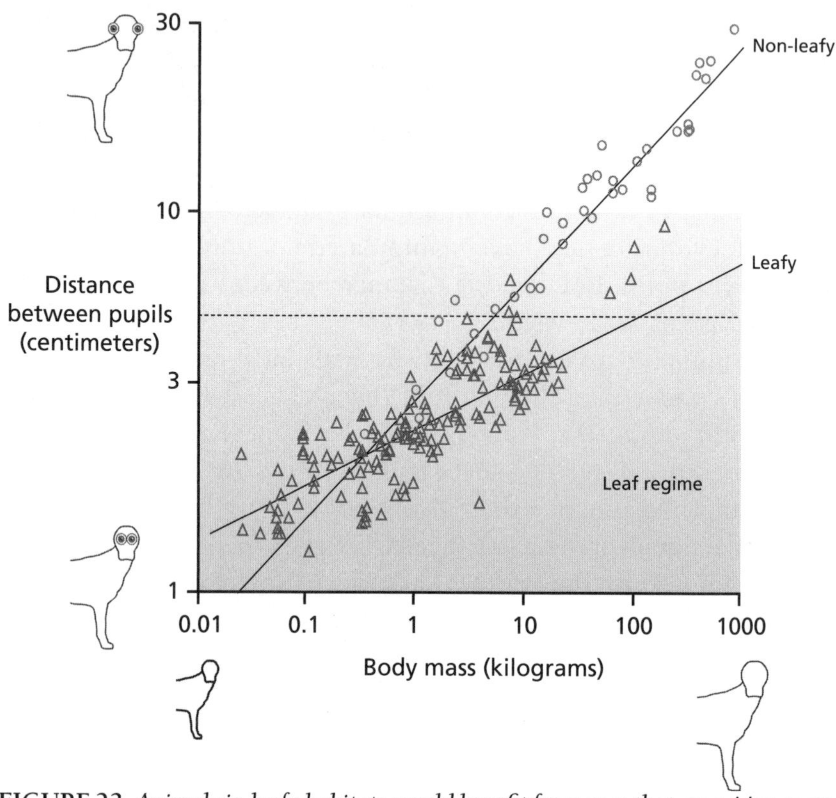

FIGURE 23. *Animals in leafy habitats would benefit from eyes that are wider apart for better X-ray capabilities, but there is no need for the distance to be much wider than that of the largest leaves. That's why we expect small, leafy animals to have eyes that are farther apart than the eyes of similarly sized non-leafy animals. We also expect large leafy animals to have eyes that are closer together than the eyes of their non-leafy counterparts, with a distance not much larger than the width of the largest leaves. We can see this in the plot of eye separation versus body size data for leafy and non-leafy animals. The dotted line is the average deciduous leaf size taken from John White and David More's* Illustrated Encyclopedia of Trees. *The gray area represents leaf sizes occurring in nature. The gray above the dotted line goes up to one standard deviation above the mean in order to give an indication of the range of larger-than-average leaf sizes. All the regions below the average are shown as gray because forests contain not just the largest leaf-size, but plant and leaf sizes below that size (including young leaves, thin branches, and so on). Note that the leafy animals, but not the non-leafy animals, seem to cluster within the range of leaf sizes found in forests. These data are from the same 319 species in Dr. Heesy's work, minus data from the semi-leafy animals, which are not shown here (they fall in between).*

second layer (because anything seen beyond the first layer of clutter tends to be seen by only one eye, so you'd have no binocular vision at the second layer of clutter). Sure, if we cluttered the world we'd have a big edge over visitors with sideways-eyes, but why ruin our vision just to make tourist fish feel inferior? Is there a way of adding clutter to our modern world that actually allows us to see *better*—which is to say, more—than we currently do? (See Figure 21c.) Can we find some way of changing the world so that we fully utilize our ability to see *two* layers in any given direction? The trick is to find a way of adding *informative* clutter, and only in those scenarios where it doesn't hinder our view of anything. Good candidates for places to add clutter would be anywhere in front of something flat and opaque, like a photograph, bulletin board, television, or computer. If you place informative clutter in front of *these*, then the clutter doesn't take anything away from your vision because, although clutter does tend to eliminate your ability to see in stereo 3-D, flat displays are, well, flat. You don't need to see them in stereo 3-D, because stereo 3-D doesn't give you any extra information about them. And even though clutter reduces your ability to see through any further clutter, flat displays like computer screens have no gaps in them to see through, so you wouldn't be able to see anything beyond them anyway. If you could find a way to put informative clutter in front of flat paper or electronic displays, then you might find a way to make your otherwise obsolete X-ray vision pay off.

Let's start with a bulletin board, where you have lots of information arrayed over a flat surface. There is no reason you need to see that array of information in 3-D, and there is also no way to see through it. Only one eye is needed; our other eye is wasted. How do we give that other eye something to do? Try bringing a bush to work, and pinning or sticking your bulletin board material on its leaves. Make sure to use a bush with multiple layers of leaves, and place post-its on the inner leaves as well as on the outer ones. If you work in a lush technology park and own a shovel, then you can experiment with many different bushes until you find one that's just right—one with Post-it-note® sized leaves, and one that is not too bushy, too sparse, or too thorny. You can arrange the notes so that they all face the same direction, or face them outward so that you have to walk around the tree to see everything. Either way, any single view of this “bulletin bush” will provide multiple layers of information, helping put your X-ray power to good use (and providing a great conversation piece).

Another idea would be to develop a long, thin, vertical computer screen to be placed in front of your usual computer screen so that you could see through it with your X-ray power, akin to seeing through a pen held vertically in front of you. Unlike with a pen, though, information could be displayed on the thin screen, giving you access to this extra information without obscuring your view of the regular computer screen beyond it. This could be crucial for workers like stockbrokers or airline pilots who have trouble fitting all the information they need into their visual field. A thin, vertical screen hardly utilizes your *full* X-ray potential, but it could be a valuable addition. Imagine stock prices flowing down the vertical screen, colors flashing whenever they need to attract your attention and cueing you to focus temporarily on it rather than the information on the screen behind it. Alternatively, you could use a pair of flat screens arranged in a wedge shape so that the right eye could see the screen on the right side of the wedge but not the left, and the left eye could see the left side of the screen but not the right. To visualize this, picture an “8” followed by a less-than sign:

8 <

The two circles of the “8” represent your eyes, and the “<” represents the two computer screens meeting at a point near your nose. This makes use of the fact that what we see when looking through something is not two *copies* of what we are seeing through, but two *views* of it. Using such a wedge would double the information available to your eyes when looking at a single vertical screen, although you would no longer be able to look directly at either screen and make it opaque, the way you would with a single screen. (The screens would also vary in depth, which might make it hard to keep them in focus.) Whether you try using a single flat strip or a wedge, if one works, why stop there? We can add several screens, styled like the fence in Figure 15, thereby utilizing our X-ray vision even more fully.

Perhaps one day we'll be able to see through traffic on the way to work, then put on our snail-like eye-stalks and see through the people around us as we walk through the crowded streets to our office. We'll leave our coats and eye-stalks at the office door, head over to our bulletin bush to remind ourselves of the day's tasks, and sit down in front of a cluttered computer screen to work. We'll see. I can only hope that we find many more ways of resurrecting the usefulness of our lost X-ray power in the future.

See Through

If eyes were cheap, we could have two eyes facing forward, two eyes facing backward, and while we're at it, two on each side of our head. We would then not only have panoramic vision, but all parts of our panoramic vision would be equipped with X-ray capabilities. Some arthropods have gotten into the spirit of this: some spiders have eight eyes, with pairs on the front and sides, and horseshoe crabs have ten eyes, located in front, in back, and along the tail, some of them paired and some of them unpaired in cyclopean style.

However, our eyes, and particularly the neural tissue to support them, are not cheap—and, at any rate, evolution doesn't appear to have had any success creating mammals with more than two eyes. We have therefore been stuck with a zero-sum game where an animal that wanted two eyes pointing in the same direction had to accept a loss of vision altogether in another direction in exchange. What advantages could there be in having two eyes pointed in the same direction that could ever outweigh the utter loss of vision elsewhere? At first glance, one might expect that there are none, since any mere incremental advantage up front must surely be outweighed by an effectively infinite loss in visibility behind. And *that's* where the power of X-ray vision is so helpful in understanding our forward-facing eyes. As we have seen in this chapter, X-ray vision allows us to recognize more objects in the world than if we had panoramic vision—but only *if* we live in clutter with leaves smaller than the distance between our eyes (i.e., if we are large and leafy-loving). X-ray vision more than compensates for posterior blindness because it adds a new *layer* of vision in front of us. And having two layers of perception in the same direction is like two lasers which together have the power to set alight your view of the world, but separately do not.

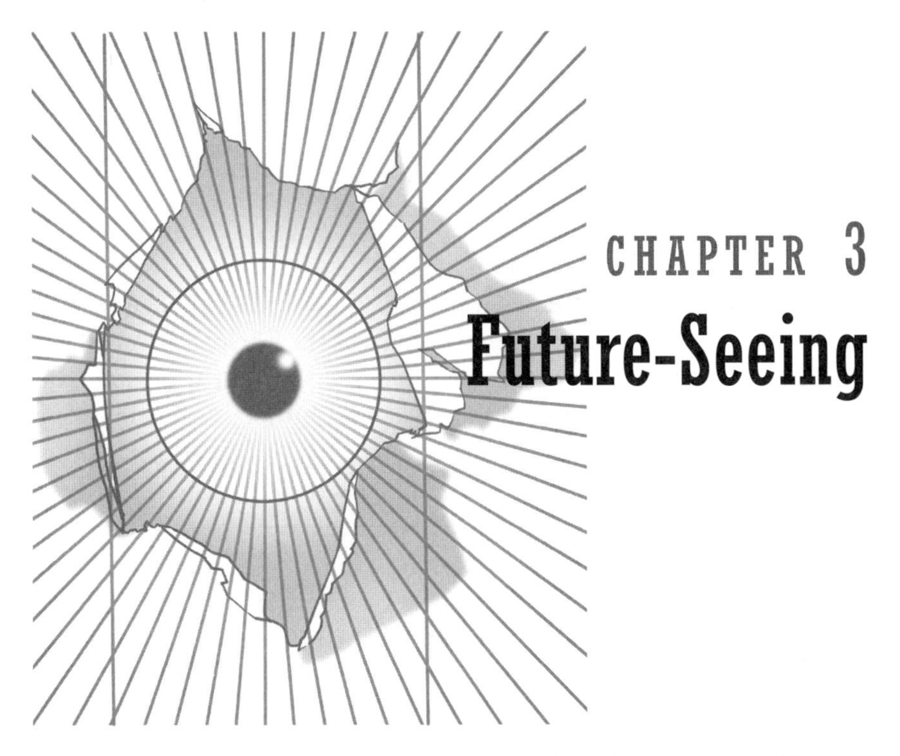

CHAPTER 3 Future-Seeing

Off ran Dingo—Yellow-Dog Dingo—always hungry, grinning like a coal-scuttle—ran after Kangaroo.

Off went the proud Kangaroo on his four little legs like a bunny.

This, O Beloved of mine, ends the first part of the tale!

He ran through the desert; he ran through the mountains; he ran through the salt-pans; he ran through the reed-beds; he ran through the blue gums; he ran through the spinifex, he ran till his front legs ached.

He had to!

Still ran Dingo—Yellow-Dog Dingo—always hungry, grinning like a rat-trap, never getting nearer, never getting farther—ran after Kangaroo.

He had to!

—Rudyard Kipling,
"The Sing-Song of Old Man Kangaroo" (*Just So Stories*)

Crystal Eye Ball

Glass spheres have long been thought to have magical powers, including giving people the ability to see the future. We know them more quaintly as "crystal balls." Figure 1a depicts a typical crystal ball, this one sitting in front of a series of screens. As you can see in the picture, the ball reveals a distorted image of the scene beyond. I don't know why glass spheres were ever thought to be magical; they are nice to look at, and those that are about the size of a baseball can be thrown farther than expected. But I suspect that if crystal balls had grown on trees or littered the creeks of Europe, the ancient druids would have chosen something else as a tool for their divinations. Perfect clarity and roundness were not easy to achieve in those days, however, so the rarity of crystal balls must have fed into the magical furor. Today you can buy one online for $24.99, and excitement about them is low. I could find no user reviews of crystal ball products on Amazon, whereas my first toothbrush selection had two user reviews (one with the heading, "Awesome toothbrushes," the other, "The only toothbrushes I'll ever use!").

While crystal balls aren't as awesome as we once thought, new research suggests that there are "real" crystal balls in our very own eyes. We have evolved eyes and visual systems with the ability to see—literally see—the future. This is not some special power we bring out only on occasion. No, this future-seeing power pervades nearly all our activities. And not just for us humans, but also for our primate ancestors, and probably for most vertebrates.

Future-seeing may sound far-fetched at first blush, but consider future-*thinking*. You have thoughts about the future all the time. Some turn out to be false, but many turn out to be true. Right now I am thinking that my infant son will wake up at 7 A.M. tomorrow morning—probably a true thought about the future. There's nothing fishy about future-thinking, then. And visual perception is just a special variety of mental process, one that leads to seeing rather than sentences running through your mind. So the idea of future-seeing isn't crazy so long as we understand it to mean the visual system creating a perception that represents the way the world will look in the future—and, of course, so long as we recognize that the brain may get it wrong.

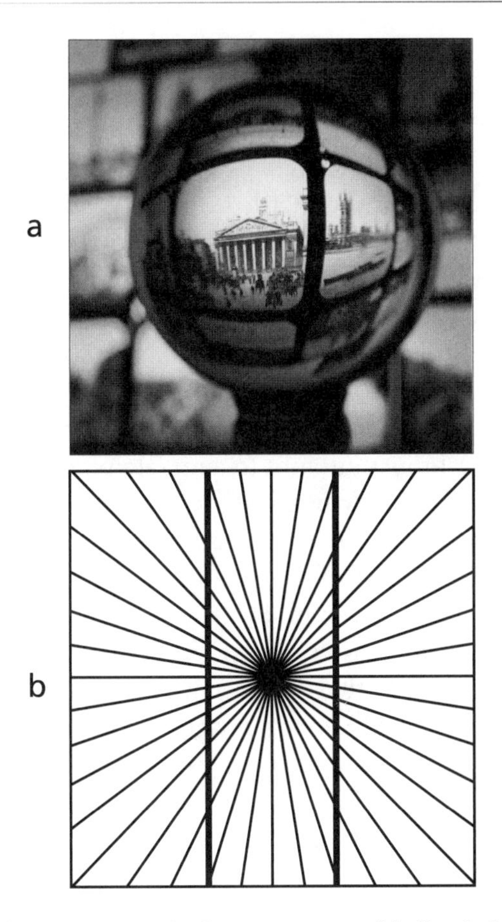

FIGURE 1. *(***a***) Trying to see the future in a crystal ball only leads to a distorted, upside-down, mirror view of the present, and persistent dizziness. (***b***) But you* can *see the future in drawings like this Hering illusion. Evolution has seen to it that drawings like this elicit premonitions of the near future.*

But "not crazy" hardly implies "plausible," much less "true." Where are these supposed crystal balls that grant us visions of the future? In our eyes and brains, of course, but little good will come of just staring at them. You can only use your crystal-ball visual system yourself, from a first-person perspective, and only by looking at the right kind of drawing or image. Looking at Figure 1b, for instance, you can see the effects of your future-seeing powers. What you perceive is different from what is actually there. Specifically, you perceive the two vertical lines to bow out at the center—even though they are actually perfectly vertical. What you are perceiving is a premonition, not present reality! Some readers might now be asking, "*This* is seeing the future!?" Well, what were you expecting? A vision of alien ships strafing Park Avenue? While we can't see the future in the sense of knowing how far away our parking space will be tomorrow, drawings such as the one in Figure 1b really do elicit perceptions of the future.

As we will see in this chapter, the story of future-seeing is entwined with the mystery of why we perceive illusions. Contrary to the established view on the nature of these illusions—which supposes that they are due to our brain's attempt to generate perceptions appropriate to a three-dimensional world, but are in error because the stimuli are on flat pieces of paper—I argue that these illusions are due to our brain's attempt to see the future, generating perceptions that are in accord with the present. Our brain believes the stimuli from these images are in fact *dynamic* stimuli, and so generates the appropriate perceptions—perceptions that are mistaken only because the stimuli are actually statically printed on the page. By the end of this chapter we'll see that my future-seeing hypothesis is able to provide a kind of "grand unified theory" for illusions, an argument I first made in publications in the journals *Perception* and *Cognitive Science*.

Please Excuse My Evolution

Eyes date back to more than half a billion years ago. If you had the chance to talk to your ancestors, once they stopped yapping about how hard things were in their day, you might casually mention that we are subject to certain visual misperceptions similar to the one in Figure 1b, or the striking ones in Figure 2. These drawings could not be much simpler—they consist of just a handful of straight lines—but even with our millions of extra years of evolution, we still can't see them correctly. One can only imagine your ancestor's disappointment.

You might try to assuage your guilt by telling your ancestor that these illusions are just silly drawings, that you're designed for the real three-dimensional world and don't make these mistakes in real life. But the illusions work in three-dimensional scenes as well. For ex-

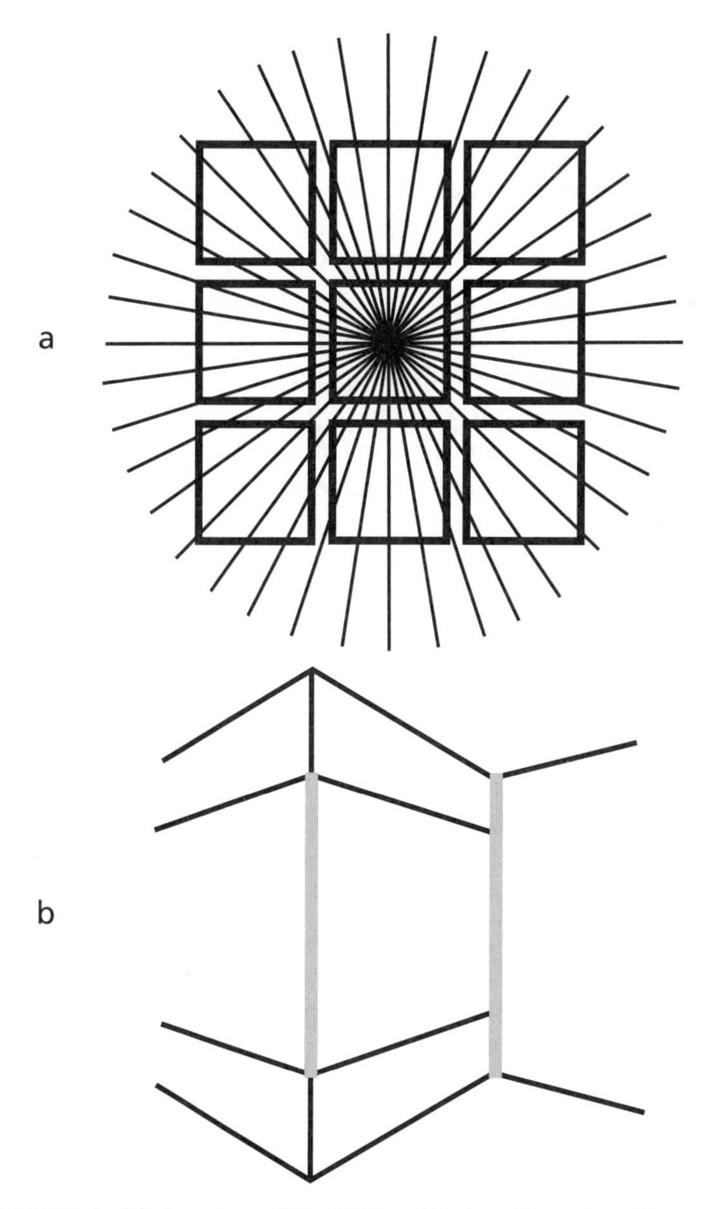

FIGURE 2. (**a**) *A variant of the Orbison illusion (due originally to Ehrenstein). The squares appear distorted, bulging outward, but actually are not. This drawing relies on the same misperceptions as in Figure 1b, but the grid here helps more thoroughly capture how our misperceptions vary within the radial display.* (**b**) *A variant of the Müller-Lyer and Ponzo illusions. The two gray vertical bars are actually the same length.*

ample, the same illusion we see in Figure 1b would also occur if the two vertical lines were two poles in front of a contour-lined hallway, because your visual system presumes that drawn images *are* from three-dimensional scenes (that is, after all, what your visual system is designed for). Figure 2b shows a case that looks much more like a hallway. If you were really in front of such a scene, you would still perceive the gray bar on the right to extend both higher and lower in your visual field than the bar on the left. Disgusted that you really are subject to serious illusions, you shuffle away from your ancestor mumbling something about a meeting and overhear your ancestor asking to be awoken again in another half billion years.

Why do we see such illusions? Some might be content to say that such misperceptions are simply mistakes—a sign of imperfect design. Evolution does not lead to perfect optimization, and that's all there is to it. Still, asking for the ability to correctly perceive a simple configuration of straight lines is not asking for some pie-in-the-sky, perfectly optimized brain. There is no reason to believe that there should be anything remotely tricky about visually processing such simple drawings. Are we to believe that the evolutionary forces, capable of creating the diversity and complexity of animals such as sea stars, squids, spiders, stingrays, and skunks, can't create a visual system that can correctly see a few lines? If you find yourself thinking such thoughts, I advise turning yourself around and rethinking the matter, for in the absence of a really good argument, with evidence, for bad evolutionary design you should presume instead that you just haven't yet been clever enough to figure out what about the design is so good. In fact, cases of seemingly bad design are usually cases ripe for big discoveries.

We clearly suffer from misperceptions when we look at the illusions in Figures 1 and 2. But as we discussed earlier in the book, our misperceptions may have a purpose: they may not be designed to truthfully capture, or "photograph," what's out there, but rather influence our behavior in a manner best suited for a particular situation. The illusions in Figures 1 and 2 could be useful fictions. In order to believe this, however, we would need to have a strong argument, and evidence that supports it; although fictions can, in principle, be useful, fact has a tendency to be more useful. When presented with a tree, say, it is typically best to see a tree, not some rainbow swirl, if your brain is hoping to prompt the right behavioral response. Therefore, while useful fictions are still a possibility, I don't see any reason to believe this is the case for illusions like those in Figures 1 and 2.

So far we have determined that these illusions are probably not the result of imperfect design, nor are they likely to be useful fictions. What else might they be? There is no getting around the fact that what you see when you look at Figures 1 and 2 are mistakes: what you see does not agree with reality. But seeing them that way need not be a mistake due to imperfect design. Instead, I will argue that when we encounter visual images like this in real life, out in the world, we are in motion, and that what are misperceptions in the case of the static drawings above would actually be accurate perceptions in real life. So in my view we make perceptual mistakes only when presented with stimuli that are unnatural (in this case, stimuli that are naturally encountered when in motion, not when stationary). In the case of illusions, our brains carry out the construction of perceptions that do not agree with what is really out there. But these perceptions are created because in natural circumstances they *do* tend to lead to true representations about the outside world. We'll see why as we proceed through this chapter. Motion is crucial to the story of illusions, and so motion is where we turn next.

Dulling Time's Blade

The present is a thin knife's edge, constantly sweeping along. The most important information about the present lies in the present, but by the time you can make sense of that information, it is, of course, too late to use it in the present, because the information is by then about the past; a new present has already arrived. How, then, do we deal with the problem of living in the present, but not looking stupid? How do we avoid bumping into things in the world; how do we move before the lion pounces; how do we tell a clever joke at a dinner party before the conversation has moved on?

One way to deal with this issue is to live only in a highly static environment. If the world never or hardly ever changes, then time becomes less like a knife's edge and more like a spacious dining room table. Or, if your environment does have dynamically changing features, make sure you don't need to care about them. Plants have taken this route, and their risk of looking stupid is accordingly much reduced.

But what to do if you live in a dynamic environment, or if you yourself are dynamic—that is, if you yourself move? A great solution to this problem would be to freeze time for everyone but yourself. You could then sit down and think hard about what to do, and start time again whenever you're ready. Bill Murray's character in the now-classic *Groundhog Day* does essentially this, but rather than choosing to freeze time, he and everyone else are stuck repeating the same day over and over again, while he is the only one that remembers he's lived that day before. By the end of the movie, he has almost God-like powers.

Forcing others into strange time loops is science fiction, so let's imagine another idea instead. Suppose you have the ability to foresee what will happen in your vicinity ten minutes in the future. You then have ten minutes to plan what to do at that future moment, allowing your witty jokes to be perfectly timed at that dinner party. This avoids the science fiction control of time, but foreseeing the future still appears no more down to earth... except that knowing what is going to happen later on and acting on that knowledge *is* quite down to earth; we do it all the time. I'm not referring here to vision or visual illusions and the issue of seeing the future. No, I'm talking about the simple fact that we often plan or rehearse what we're going to do before we do it. If you know you are giving a presentation in a week, then you plan for the presentation during that next week. Nothing spooky about that. Knowing the future, then, can be as powerful as controlling time, and is much easier to actually accomplish. That is, by knowing the future, you can anticipate it, plan for it, and act intelligently when it arrives. *This* is how animals deal with the knife's edge of time.

In this light, consider visual perception. We and our vision-possessing animal brethren—such as Kangaroo and Yellow-Dog Dingo from the Kipling quote at the start of the chapter—are certainly not vegetables in a static world. We run, hop, slither, swim, burrow, swing, and fly through our world, and our vision must accordingly deal with the razor-thin present. When the visual system generates a perception at time t, it would be advantageous if that perception accurately represented reality at time t. *That* would be an "intelligent" perception to have. But how can this be done, given that by the time the perception is built from the retinal information available at time t, time t will be in the past? That is, light reaches your eye at time t_1, but your brain builds a perception of it at a later time, t_2. Therefore, your perception of what the world is like at t_1 does not actually occur until t_2. Your perception ends up being about the past rather than the present.

In the discussion above, we concluded the obvious: If you know the future, then you can prepare for it and act intelligently when the future arrives. If you know (or your visual system knows) what will happen in the future, your brain can prepare an intelligently built visual perception appropriate for it in time for it to arrive. Namely, your brain can construct a visual perception informing you of what's occurring at *that* moment rather than the moment just past. If your brain does this successfully, you will perceive the *present* (as in Figure 3b) rather than the recent past (like in Figure 3a). In other words, when light reaches your eyes at time t_1, your brain should not create a perception of what's happening at time t_1, but instead create a perception of what *will* probably be happening at t_2—the time that the perception is complete. You need to foresee the future in order to perceive the present!

Are our brains really so slow that we must resort to future-seeing? How long does it take for our visual systems to take light and turn it into a visual perception? The answer is about a tenth of a second. If we simply generated a perception of what's out there when light hit the retina, then by the time the perception itself occurred, it would represent the way the world *was* a tenth of a second before. If a tenth of a second sounds too small to be worth troubling over, consider simply walking. Even if you were walking at a painfully slow one meter per second, you would be moving a tenth of a meter—or ten centimeters—in a tenth of a second. If your brain didn't bother with future seeing then when you perceive an object to be within ten centimeters of passing you, that object would in actuality have already passed you,

FIGURE 3. (a) *A person trying to catch a ball without the ability to foresee the future and thus perceive the present. When her perception indicates the ball has not yet reached her, the ball has actually just bumped into her.* (b) *A person able to foresee the future will have the same perception as in (a), but have it when the ball is actually at the perceived distance. Therefore, she will be able to catch the ball.*

or bumped into you. The discrepancy becomes considerably worse if you are running. And consider catching a ball thrown to you at a speed of ten meters per second, or about twenty miles per hour. Without using future-seeing to compensate for your brain delay, by the time you perceive the ball at a reachable meter out in front of you, the ball would have moved one meter beyond that point in space, and about twenty-five degrees beyond that point in your visual field.

One can see this future-seeing ability in a simple experiment invented by vision scientist Dr. Romi Nijhawan, now at the University of Sussex, during the 1990s. In the experiment, a light bulb is flashed just as a ball moves past it. This is illustrated in Figure 4a. Although this is what *actually* happens, we instead perceive the ball as *past* the flash, as in Figure 4b. Because the flash appears to lag behind the ball, this perceptual effect was named the "flash-lag effect." And vision scientists Professor David Alais of the University of Sydney and Professor David Burr of Florence University have shown that this works even if the moving object or the flash, or both, are not seen, but only heard.

Why does this happen? To perceive the present, we must be able to perceive (at time t) the true position (at time t) of the moving ball, as illustrated in Figure 3b. The moving ball is predictable, and the brain can extrapolate where the ball will be in a tenth of a second. The flash, however, is unpredictable, and so our perception of it occurs a tenth of a second after it happens. But in that amount of time, the moving ball has moved past the flash. This means that, by the time you perceive the flash, your perception is that the moving ball is *past* the flash (Figure 4b).

A similar effect was noticed by the vision scientist Dr. Donald MacKay in 1958, although its relevance to the timing of perception was not appreciated at the time. You can duplicate his experiment the next time you're at a disco dance. Bring with you a lit torch (although a cigarette or flashlight may also suffice). Wait until the strobe light begins during "Night Fever," and then lurch the torch wildly about. What you will notice, and what MacKay noticed presumably by different means (namely, in a laboratory with an oscilloscope), is that the handle of your torch—which is only visible during the strobe flash—appears to be visually separated from the lit part of the torch. The handle appears to lag behind the fire! (Fortunately, the illusion also puts you at an advantage in outrunning scientifically uncurious bouncers.) This works just like the flash-lag effect; here, the fire is the moving object (because it is always visible), and the handle is like the flash (because it is only intermittently visible, when the strobe light flashes).

The flash-lag effect concerns *where* within your visual field you perceive the moving object to be relative to the flashing one, but what about the objects' other properties? For example, suppose an object is sitting still, but its color quickly changes from red to blue. And suppose that, just when it is a certain shade of purple, a nearby uncolored object briefly flashes the same purple color. What color do you perceive the color-changing object to have when the other

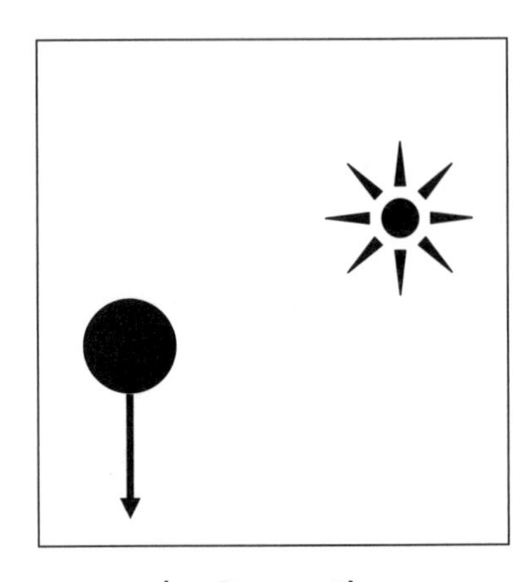

FIGURE 4. *(a) What actually happens in the flash-lag experiment originated by Romi Nijhawan. The flash occurs just as a moving object passes it.* (**b**) *How you perceive the position of the object when the flash occurs. You perceive the flash to occur after the object has already moved past. Nijhawan suggested the natural explanation here: that we are perceiving the present for the moving object, but not the flash, so that when the perception of the flash finally occurs (a tenth of a second after the object passes the flash), the perception of the moving object is in accord with the new position of the moving object, slightly past the flash.*

object flashes that purple? Bhavin Sheth, Romi Nijhawan, and Shinsuke Shimojo of Caltech showed that, just as in the flash-lag illusion, the flashed-purple appears to lag behind the changing-color object. That is, even though the uncolored object flashes purple just when the color-changing object is exactly that color, observers perceive the color-changing object's color to have already passed that purple and become more blue. Sheth, Nijhawan, and Shimojo found the same effect for a variety of other properties, such as brightness and even the pattern painted on the objects.

The uncompromising crispness of time has demanded that we evolve future-seeing capabilities, for only by seeing the future can we perceive the present rather than the past. These flash-lag effects were some of the first pieces of evidence supporting our ability to accurately perceive the present, but they were just the beginning. In the rest of this chapter I'll unveil a variety of other evidence that we have this power, and explain how the illusions we saw earlier (like in Figures 1 and 2) are a result of this power. But I don't want to leave the impression that our future-seeing power is a burden thrust upon us by the problems of dealing with ever-moving time. Instead, we should look at it as an opportunity. It's this that we turn to next.

My Supercomputer Is Running Slowly

"Is this computer fast?" I asked the salesman.

"Oh, yes," he responded.

"Mind if we start it up?" I asked.

"Sure thing." And he set the hard drive whirling.

A minute and a half later, the computer was up and running. *Awfully long for a fast computer*, I thought. *Didn't my computer take less time to boot up in the 1980s?*

Computer designers want their software to do nifty things, and the longer the designers can get the user to wait, the niftier things the software can do in the interim. But people are only willing to put up with so much waiting, and it appears computer designers are aware of that. As faster computers have arrived, computer designers have not appeared to conclude, "Now we can compute the same old things nearly instantaneously!" Instead, they seem to have thought, "Now think about how much more we can compute while the user waits!"

Come to think of it, kids work pretty much the same way. It takes a long time for a three-year-old to get herself ready for a bath; just removing her socks may take a full minute. But my daughter is five now, and I've noticed that even though she's capable of removing her clothes lickety-split, she still takes just as long (or longer) to finally join her baby brother in the bath. Rather than using her enhanced motor skills to get into the bath faster, she uses the time she used to take—the time she knows she can get away with—to carry out more and more shenanigans.

I suspect this is also the case with vision. We animals want, and evolution selects for, visual systems capable of highly complex mental calculations that ultimately end in an effective visual perception. But the brain, like the software engineer, would like as much time as possible to carry out these difficult calculations. The question is, how much time can the brain "get away" with to compute a visual perception? The answer to this depends mostly on the environment, and how predictable it is. Predictability will differ from environment to environment, e.g., from the unpredictable New York Stock Exchange to the predictable back-and-forth hits of a tennis match. In more predictable environments, the brain can accurately predict farther into the future, and thereby buy itself more time for fancier visual perceptions. That is, just like computer engineers and five-year-old bath-takers, the brain may well be designed not to minimize the time taken to build a perception, but instead to maximize the delay (by increasing its future-seeing powers) in order to secure more computational power. More generally, our brains may be designed to optimally balance the benefits of longer computation time with the greater difficulty of future-seeing farther into the future.

That is, what I am emphasizing here is that perceiving the present is not necessarily a power we possess in order to deal with some fixed neural delay. In my academic work, which has to be very concise, I gloss over the subtle point I am trying to make here, simply saying that perceiving the present is advantageous for an animal because it compensates for neural delays from retina to perception. But this can be misread to imply that the length of the delay is simply the bare minimum of time required for signals to travel from the retina to the appropriate brain areas and for some requisite computations to be done. It is, however, more complicated than this. The neural delay is itself selected for by evolution, and may be much longer than if it had been selected to be as short as possible. We are probably designed to have much longer neural delays than "necessary," giving us extra time to build better perceptions. But this is a strategy that is only made possible because we have harnessed the predictability found in nature. Counter-intuitively, then, it may well be that the slower-to-react brains are the smarter ones.

Perhaps lurking beneath our musings on the brain's reaction speed is an explanation for Spider-Man's spider sense, with which he senses impending danger—often danger directed at him. Because he is able to anticipate events several seconds before they happen, he has more time to prepare appropriate perceptions and behavioral responses for when those events finally occur. In order to see so far into the future, Spider-Man must be tapping into regularities in nature that we haven't noticed—which seems reasonable, given that the nasty spider bite may have given him enhanced senses. If neurophysiologists were to measure the time it takes Spider-Man to generate a final perception from his incoming sensory stimuli, it may be considerably longer than our own delay. This would not be a sign of a low intelligence, but instead a result of his greater power to foresee the future, because it would allow his brain more computation time for building better perceptions of the present. It would actually be a sign of Spider-Man's genius.

The Moment That Never Comes

Earlier I mentioned how you can make yourself look smart either by controlling time itself or by foreseeing the future. The problem with the foreseeing-the-future strategy, however, is that if you're wrong, then you really look the fool. Psychologists love to make people look like fools—it may be why most of them go into the field—and several psychologists have discovered compelling cases where your visual system is tricked into thinking something is going to happen, but then it doesn't. That is, they have found ways to trick your visual system into perceiving the next moment while ensuring that anticipated next moment doesn't arrive.

One class of such tricks is called "representational momentum," discovered by University of Oregon psychologist Jennifer Freyd in the 1980s. In her original experiments, which have since been much elaborated upon by researcher Timothy L. Hubbard and others, she showed people simple movies of a rotating rectangle like the one on the left of Figure 5, and asked them what they remembered the final frame to look like. She found that people falsely remembered the final frame to be further along the rotation than it actually was in reality.

Vision scientists Vilayanur Ramachandran and Stuart Anstis of UC San Diego in 1990, as well as Russell and Karen De Valois of UC Berkeley in 1991, discovered a second trick, an effect related to what is called "motion capture," where the motion within an object

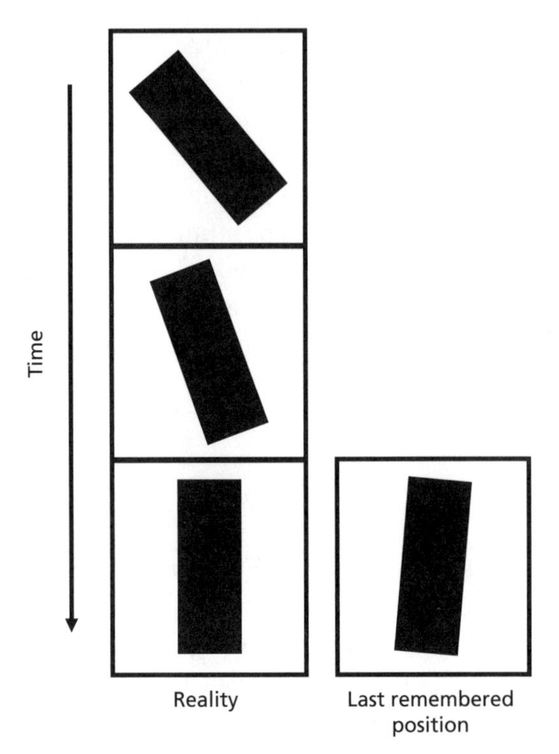

FIGURE 5. *An example representational momentum experiment done by Jennifer Freyd in the 1980s, in which a rectangle is shown rotating, as seen in the sequence of frames on the left. People remember the rectangle's final position to be further along in the rotation than it actually was—which is consistent with perceptual premonitions.*

suggests the object is going to move in a particular direction, but it never does. Observers perceive the object to be shifted in the direction it "should have" moved (Figure 6).

As a final example, consider the image shown in Figure 7. Each of the objects has a blur trailing behind it consistent with a moving object. Many observers perceive these objects to be moving away from their blur-tails, with the longer-tailed dots appearing to move more quickly. In other words, people perceive movement in these static images—movement consistent with the direction and length of each blur. (We'll take up the blur concept in more detail later in this chapter.)

All these effects are similar in that people perceive objects to be where they "ought" to be, but because the images don't change like

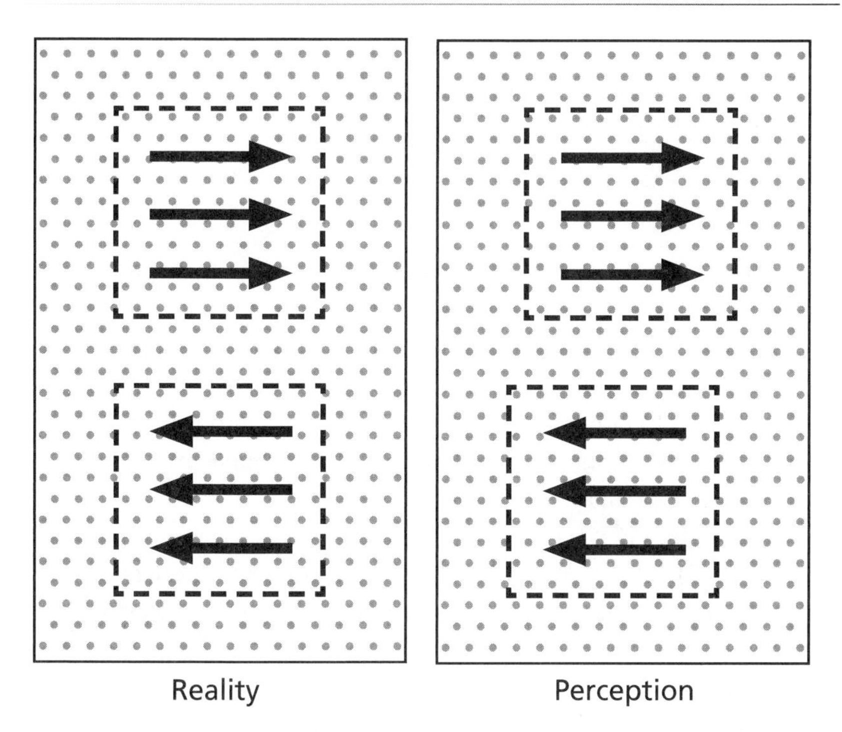

FIGURE 6. *An illustration of an experiment in which the dots inside the "squares" move in the indicated direction. In the experiment, there are no contours surrounding the squares; your perception of the squares is due solely to the fact that the dots inside the region are moving and the dots outside the region are not. The top square appears shifted to the right of the lower square because the dots are moving toward the right. Experiments of this kind were carried out in the early 1990s by Ramachandran and Anstis, and by De Valois and De Valois, who each suggested that the perceptions were a result of the visual system's attempts to perceive the present (by creating a perceptual premonition).*

they "should," a misperception is the result. That is, our visual systems foresee the future and use that information to generate a perception of the present, but the foreseen future never arrives, leading to an illusion. This is the kind of misperception we'd expect to see if we perceived the present, and in fact, the researchers involved in the second effect mentioned above were the first to propose that our visual systems might be trying to perceive the present. As we will see, the enigmatic geometrical illusions in Figures 1b and 2 are actually a lot like the cases in Figures 5, 6, and 7, and are especially similar to the image in Figure 7 (in that, like them, it is static). The geometrical illusions, however, are bound up with the fact that animals move forward, something we'll take up later on.

Masking the Tape

Being fooled by the illusions in Figures 5–7 may be embarrassing, but lucky for us, experimental psychologists do not run reality. In the real world, moments *do* arrive. In the real world, the future is not put on pause and our perceptual predictions are not left waiting for the expected. In the real world, our perfect powers of future-seeing are always correct, and so we therefore always have an accurate perception of the present.

Whoa there! "Perfect powers of future-seeing?" I was just checking to see if you were awake. Unfortunately, we don't have *perfect* powers of future-seeing. It's just not possible. And if our future-seeing sometimes fails, it means our perceptions of the present sometimes fail. Furthermore, these failures are likely to be even *more* embarrassing than simply perceiving a future that never arrives. In reality, it isn't that objects that are expected to move forward don't. Instead, objects that "should" go one way end up going in some unanticipated direction. That is, we don't just falsely perceive something to be there when it is not there. Rather, we perceive an object to be one place when it is actually somewhere else—say, ricocheting dangerously near our head. Perceptions of the present can at times be perilously ignorant of the present.

Is this a problem for future-seeing? Well, it's a problem that has not been *overcome* by future-seeing, but it is a not a problem *for* future-seeing, per se. After all, if an animal did not try to perceive the present, while it would never have a perception of a falsely anticipated present, it would also never have *any* perceptions of the present, accurate or otherwise. When the dangerously ricocheting object careens toward the non-future-seeing animal's head, he too is blissfully ignorant—not because he is perceiving a different or false present, but because he is still busy perceiving the way the world was in the past before the unanticipated ricocheting began. If you have to be ignorant of the present, you might as well be fantasizing about the present you hoped for, rather than dwelling on the past.

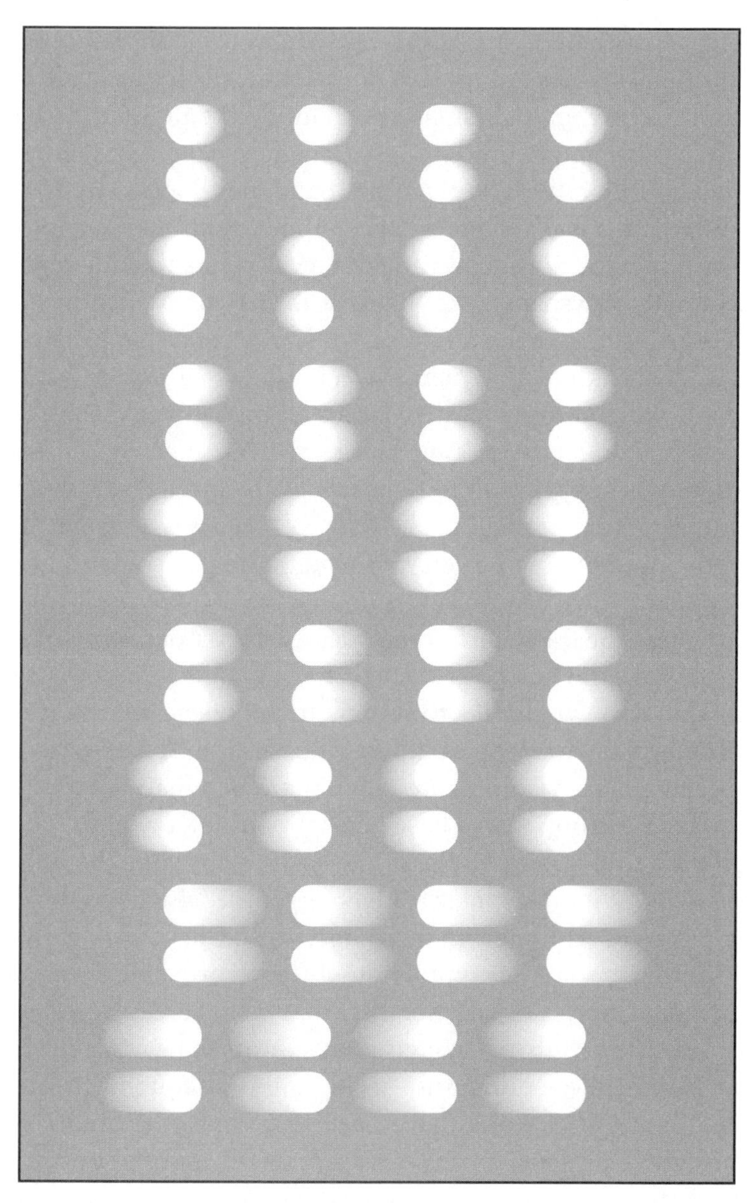

FIGURE 7. *An illusion in which many people see movement consistent with our visual system's attempts to perceive the present. The objects appear to be moving away from their blurred tails, and the longer-tailed objects appear to be moving more quickly. This illusion shares some similarities with works by the artist Akiyoshi Kitaoka, "Hearts and Tears" in particular. (For some people, the illusion is enhanced if you hold it at arm's length.)*

Except that there is one difference between the incorrect future-seeing animal and the animal that does not see the future at all. When the future-seeing animal fantasizes about a false present, he has recorded a false memory onto his visual memory tape. The animal that does not see the future, on the other hand, never adds anything but the truth to his visual memory tape, although this tape is somewhat useless in the knife's edge of the present. It is like the difference between a daily newspaper like the *New York Times*, which must desperately try to anticipate the results of an election so it can go to press as soon as possible, and a weekly news magazine like *Time* magazine, which can patiently wait for all the polls to dribble in before deciding on its cover. *Time* magazine lives somewhat outside of the present, is less relevant to the news of any given day, but is more likely to get its facts straight. The *New York Times* is dynamically living in the present, a more relevant read on any given day, but more prone to occasional errors. Perception is like the *New York Times*.

How does the *New York Times* deal with its occasional future-seeing mistakes? They issue a correction the following day. But we know what they would *like* to be able to do: magically replace each of their mistaken newspapers around the world with a corrected one. And because false headlines are much more humiliating for a venerable news institution than a false perception is for any of us, I'm sure that, if they could, they would use a "flash" device like the one used in *Men in Black* to make their readers forget the false news they originally received. Although their readers would have originally read false information, at the completion of these corrective shenanigans those true memories of falsehoods would be gone, replaced by false memories of the truth. Amazingly, that appears to be what our brains do for us whenever our future-seeing fails. Although our original perception at time t would have been false, once the brain realizes the mistake it "flashes" away, or masks, the memory of our original misperception and creates a new (though false) memory of having perceived the truth at time t.

Or, at least, it does sometimes. We saw earlier that there are cases of future-seeing embarrassment that our brains don't paper over. In particular, we saw several cases (Figures 5–7) where cues suggest what the next moment will be but the next moment refuses to show up.

FIGURE 8. *Newspapers that make a mistake would like to be able to change recorded history so that their embarrassing mistake never happened. They just can't. The visual system has fewer ethical guidelines, and does just this, masking true memories of false perceptions and replacing them with false memories of true perceptions.*

Professor Romi Nijhawan, the fellow we discussed earlier regarding the flash-lag illusion, and a graduate student named Gerrit W. Maus believe there is a simple reason why the brain fails to overwrite those false future-seeing perceptions: the stimuli used in those experiments are unnatural, and so the brain didn't evolve to deal with them. In real life, when the brain fails at anticipating the future it is usually because an object suddenly bounces or bumps when it should have smoothly continued doing whatever it was doing originally. That is, typically, a failure to perceive the present is due to some kind of unexpected *change*, or discontinuity, in the world. In the three illusions discussed above in Figures 5–7, however, there was no large discontinuity in the visual stimulus marking an unexpected moment. For example, in the case of Jennifer Freyd's representational momentum experiment (Figure 5a), the sequence of frames is highly discontinuous at each step to begin with (i.e., the movie was not a smoothly rotating rectangle), so the disappearance of the rectangle in the final frame is no more sudden a change than the changes that occurred in the previous moments. In the motion capture experiment (Figure 6), there is no end to the visual sequence—the motion inside the objects is never-ending. There is also no sudden change occurring in the image in Figure 7; in fact, there are no changes at all, given that it is a static image. Because there is no strong change in the object the future is being perceived for, Nijhawan and Maus believe that the brain is less able to notice that it *has* been tricked. These researchers point out that, unlike the cases in Figures 5–7 where there are no sudden changes, when the unexpected *does* happen the way it typically happens in nature—accompanied by an abrupt change—we don't have misperceptions of the future. Or rather, we *do* have misperceptions of the future, but the abrupt change signals the brain's visual tape editors to mask the false perception, writing over it with a true one we did not in fact have at the time.

That's why we don't see this kind of illusion—which researches call an overshoot illusion, because the brain is overshooting the actual spatial location—in real life. And that's why, as researchers have discovered, if you add sudden changes to an overshoot illusion, the overshoot tends to go away. For example, recall Romi Nijhawan's flash-lag effect (Figure 4), where there is a moving object flying by, and just as it passes a darkened bulb, the bulb flashes. People perceive the moving object to have already passed by the bulb when the flash occurs, a perception consistent with perceiving the present. What happens, though, if the moving object suddenly disappears just as it passes the bulb? That is, what happens if, after the flash, both objects are gone? Our future-seeing brains should have created a perception of the object as having gone farther than it actually did (i.e., overshoot), and observers should say that they perceive the flash to lag behind the moving object, just like in the flash-lag effect. However, that is not what happens, as visual neuroscientists David Whitney and Ikuya Murakami from Harvard in 1998 and David Eagleman and Terrence Sejnowski from the Salk Institute in 2000 pointed out. Instead, observers perceive no overshoot in this case. For Romi Nijhawan, who had first noticed this in 1992, this makes sense, because the moving object's disappearance is unanticipated—there is a sudden change to the moving object. Objects disappearing is one kind of unanticipatable event we must surely have evolved to be good at dealing with, because moving objects disappear from view all the time (for example, when suddenly occluded by something). The way we deal with it is by clever tape editing after the fact. As a test of their idea about the importance of abruptness for initiating the masking process, in 2006 Nijhawan and Maus created an experiment similar to the one above where the moving object suddenly disappears, but instead, they showed observers a moving dot that fades away; no flash was involved. In one case the moving dot was more extrapolatable than the other (its movement was more predictable, by virtue of it having a longer trajectory), and Nijhawan and Maus found that observers perceived the more extrapolatable fading dot to become invisible at a later position than the less extrapolatable fading dot. That is, in these conditions—when there is no abrupt change—overshoot again occurs.

Non-abrupt unpredictable changes like these are unnatural, so the brain does not get the expected feedback that an unanticipated event has occurred, and so does not realize it has any tape-editing to do. But non-abrupt unpredictable changes are just one of many kinds of unnaturalness that have the potential to embarrass the visual system. We might expect that our brains are able to carry out embarrassment-reducing tape-editing only in natural situations, and so we might wonder what other varieties of unnaturalness we can use to foil our internal visual tape editors. Vision scientists Ryota Kanai, Bhavin Sheth, and Shinsuke Shimojo at Caltech in 2004 carried out experiments just like the flash-lag variant above, where the moving dot disappears when the flash occurs, except that they added unnatural ingredients of various kinds. For example, when observers were forced to look to the side so that the moving dot was well in their peripheral vision, they perceived the moving dot to overshoot its true final position. And when they lowered the brightness of the moving dot so that it differed only a little from the background, observers again perceived overshoot.

Another case of perceptual overshoot due to unnaturalness was carried out by the visual neuroscientists Yu-Xi Fu, Yaosong Shen, and

Yang Dan of Berkeley in 2001.They used two fuzzy balls that, positioned one above the other, moved toward the center of the screen, one from the left and the other from the right. The moment the balls aligned, they reversed direction and went back the way they came. Observers, however, perceived the fuzzy balls to have passed one another before they reversed direction—i.e., to have moved farther in their original direction than they actually did. In all of these illusions, the fundamental similarity is that they make the position of the moving object more uncertain. It's as if the *New York Times* became less able to notice their own mistakes—and less able to, in reaction, go out and flash everyone *Men in Black*–style—because their writing has become more ambiguous or vague. (If they had a literary theory section, they'd never have to issue any follow-up corrections.)

Super Salamander Vision

Few readers probably began this chapter expecting to discover that we can actually read the future, much less expecting to find a how-to manual on better tapping into our latent powers to make an easy million by working at home. But perhaps you were nevertheless disappointed that our future-seeing ability only goes a tenth of a second into the future. And perhaps you were even more disheartened to learn that by the time future-seeing has been perceived by your brain, the future has arrived, meaning you are really seeing the present and not the future at all! The only time you actually perceive the future is when the future fails to arrive, in which case it wasn't the future after all. But before you cry "bait and switch!" and get the Better Business Bureau after me, let me explain...

...that things are even worse than that. If you were thinking that this future-seeing power is not quite the same as the power fortune tellers claim to possess, but is nevertheless an important human visual power that sets us apart from the rest of the animal kingdom, then get ready to feel, uh, disillusioned. Why? Because rabbits also have this power. So do lowly salamanders have this power. Furthermore, the retinas of rabbits and salamanders can do a pretty good job at future-seeing all by themselves—without help from the rest of their puny brains. Neuroscientists Michael J. Berry, Iman H. Brivanlou, and Markus Meister of Harvard, and Thomas A. Jordan of Stanford showed in 1999 that neural activity in rabbit and salamander retinas does not lag behind the position of a moving object, but instead maintains the actual position of the moving object. Their retinas alone perceive the present via future-seeing!

Our super human vision may no longer seem quite so super after learning that. But I would argue that whether or not we humans are super should not depend on whether we can brag about it to the amphibian neighbors over dinner. What makes our power noteworthy is simply that we tend to think of the power of future-seeing as beyond our capabilities, and that we think it would be a really neat power to have, so much so that we frequently like to imagine superheroes and other fictional characters with such powers.

Furthermore, although our future-seeing power is only used for perceiving the present, not the future, would we really *want* to be perceiving the future? I want my perception at time t to accurately reflect the world at time t. *That* is what will help me make intelligent decisions and act appropriately, because my body lives in the present, not the future.

Although I have talked about a variety of misperceptions in this chapter thus far, I have not shared many illusions, nor have I discussed why we perceive them. The research we discussed above, which supports the hypothesis that our brains perceive the present was not, in fact, about illusions at all. While it is true that the experiments concern misperceptions, none of those researchers realized that there might be a connection between their work and century-old optical illusions, like those in Figures 1 and 2. This is where I enter the story. In the late 1990s, I knew nothing about flash-lag or related arguments regarding how we perceive the present. But I had noticed that I could parsimoniously explain those old, famous illusions *if* it were true that our visual systems try to perceive the present. I was elated when I found that I was not the only one making the radical-at-the-time hypothesis that our visual systems were designed to perceive the present. In the remainder of this chapter I discuss the even more alluring second half of the future-seeing story: illusions.

Illusory Illusions

There are downsides to having superpowers. For example, Superman tends to accidentally crush most of his dates, and Spider-Man has developed the bad habit of darting out the window right before his girlfriend utters the words, "So, where is this relationship going?" And the downsides are not just a problem for unlucky paramours. Consider our own Spider-sense-like power of seeing the future. We can't just sit back and wait for the world to tell us what's happening; doing so would lead only to perceptions of the past. Instead, we have to dynamically engage with our future-seeing power, in the sense that we must constantly be anticipating the next moment and building a completed perception of it by the time it arrives. We have to surf on the crest of the constantly moving present.

Surfing, however, is dangerous. As we saw earlier, it can lead to embarrassing misperceptions when the world doesn't go as anticipated: you can perceive a future that does not come to fruition. Earlier we also discussed how Spider-Man's Spider-sense might be a longer-range version of our future-seeing power. When we extrapolate incorrectly, we can be significantly wrong, but not more than a tenth of a second's worth. Any illusions resulting from mistaken future-seeing will be relatively small, although potentially still debilitating. Spider-Man, however, may have the power to future-see several *seconds* into the future, and that means that when he's mistaken, he'll see a super illusion. Rather than just perceiving a ball to overshoot its true position by a bit, Spider-Man will perceive the ball to have landed on the ground and rolled a few dozen meters, even if the ball has actually been vaporized in mid-flight by the Green Goblin seconds before. The point is that when we stick our necks out to perceive the present, we take the risk of being cut—in this case, suffering illusions.

Although we mentioned several illusions earlier in the chapter, there are actually tens of thousands of them, collected like butterflies by visual scientists over the centuries (Figure 9). In bulk, this pile of illusions can be used as "exhibit A" that our visual system is error-prone and incompetent. In light of the propensity of future-seeing to laden us with illusions—at least in the laboratory or an illusion book, if not so much in the real world—how many of these collected illusions

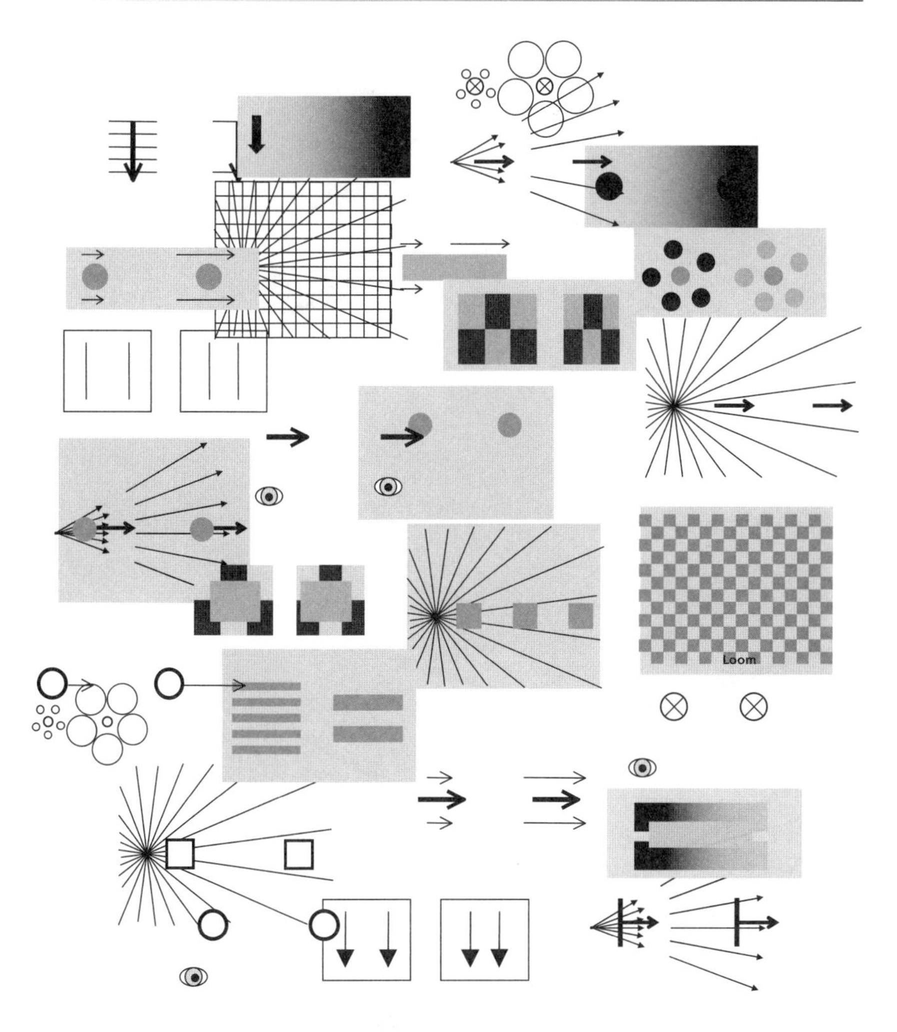

FIGURE 9. *Each icon represents a class of illusions, some of which consist of hundreds of known illusions. For example, the classical geometrical illusions from Psychology 101 courses that are also found in popular books on illusions are represented by the radial display with two squares near the bottom left. This figure is intended to illustrate the disorder among the "butterfly collections" of illusions over the years. Later in the chapter I'll describe what each icon represents, and by understanding our future-seeing power we'll be able to systematize and explain all of them (see Figure 20).*

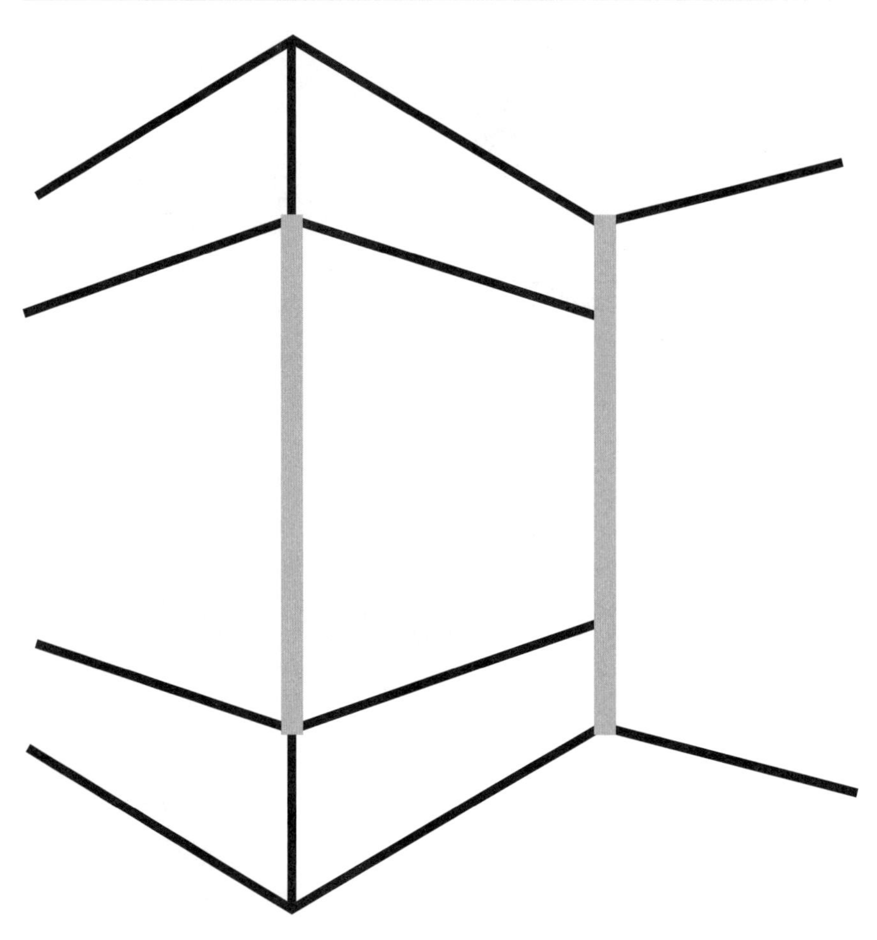

FIGURE 10. *In this drawing, the two gray vertical line segments are the same length on the page. The right vertical gray line segment appears physically longer in the depicted scene than the one on the left, but this is not an illusion, because this would be the correct perception if this were a three-dimensional scene. Illusions occur when two objects are the same in regard to a quality, X, but are perceived to differ in regard to X. You perceive the physical length of the right line to be greater than that of the other in the depicted scene, and the physical length of the right line is greater than that of the other in the depicted scene. This then is not an illusion. What is illusory about this figure is that the line segment on the right side of the depicted scene appears to fill more of your visual field than the other line segment, even though they in fact fill the same amount.*

might be due to our future-seeing powers? As we'll see, the answer is "a lot." In fact, by understanding the variety of ways in which our future-seeing might fail, it is possible to pull out hundreds of museum drawers filled with carefully pinned illusions and systematize them into a kind of "periodic table" of illusions. The troves of embarrassing evidence of our visual incompetence are, in actuality, a testament to our superpower—our future-seeing power to perceive the present.

Before we begin to systematize all these illusions, we need to pin down exactly what it is about illusions that is illusory. To start, consider the following cognitive illusion:

I splashed next to the bank. **There was a run on the bank.**

Consider the word "bank" here. The first time you read this word, you thought about river banks. The second time you read it, you thought about money banks. And yet it was the same word both times! Just mind-boggling! Well, no, not really—it's just the result of ambiguity. The definition of an illusion is when two objects that are the same in regard to a quality, X, appear to be different in regard to X. So is this really an illusion? You perceive the two instances of "bank" to differ in their meanings despite being the very same word, but the two instances of "bank" *do* actually differ in their meanings. No illusion here!

Now consider the drawing in Figure 10 (a repeat of Figure 2b), but suppose you are actually standing in the three-dimensional scene depicted rather than in front of a flat picture. This is what your brain believes it is seeing anyway; it's what your brain was designed to respond to. It perceives the gray vertical line on the left to be physically shorter (in meters) in the depicted scene than the vertical line on the right, even though the stimuli on the page (and on the retina) are the same length in each case. That is, if you were to step into the scene as it is depicted and stand beside each line segment, the one on the left would be shorter than the one on the right. Is *this* an illusion? No, and for the same reason the "bank" illusion wasn't an illusion. You perceive the two line segments to differ in their physical sizes in the depicted world, and they *do* differ in their physical sizes in the depicted world. Again, no illusion.

There tends to be a great deal of confusion about this. The standard explanation for this type of illusion is summarized nicely by the Wikipedia entry for "Ponzo illusion" at the time of this writing:

> In this context, we interpret the [right side] line as though it were farther away, so we see it as longer—a farther object would have to be longer than a nearer one for both to produce retinal images of the same size.

But as we discussed above, seeing the right line segment as physically longer in the depicted scene is not an illusion because the line segment in the depicted world *is* physically longer than the left line segment. So if there is an illusion in Figure 10, the standard explanation does not address it. It only addresses the obvious, non-illusory fact that in order for a farther-away object to be the same size on the retina, it must be physically larger.

Could there, then, be something else about the image in Figure 10 that *is* illusory? As mentioned above, the definition of an illusion is two objects that are identical in regard to X but that are nevertheless perceived to differ in respect to X. Based on that definition the natural question to ask is, "What is *identical* in the depicted scene?" The gray line segments are identical on the page, which means they each take up the same amount of space in your visual field. That, then, must be where the illusion lies: the line segments are identical in regard to how much space they take up in your visual field, but are nevertheless perceived to differ in this respect. And that is, indeed, the illusion that observers experience. Observers perceive the right line segment to extend both higher and lower in their visual field than the left line segment. That is, observers perceive the right line segment to have greater *angular* size (i.e., the amount of space filled within your visual field, or how much of the retina is covered).

Some visual scientists will say that we humans only pay attention to perceptions of physical size, and disregard perceptions of angular size. "Why would we want to perceive what's on our retina?" they ask. I'll explain that shortly, but all you have to do to see that they are wrong is move your head. Something about your perception of objects changes when you do so, even though your perception of their physical properties (like physical size) does not change. For example, if you stare up at the inside corner of your room, you will see a three-dimensional corner composed of three 90 degree angles. But if you move a few feet in any direction while still staring at it, your view changes. What you perceive changing is the "Y" shape formed by the walls as they come together at the corner: the angles you see between the lines of the "Y" change as you move throughout the room. For another example, hold up your finger and identify an object in the room a little to the left of your finger (it may help to close one eye for the moment). Shift (but don't rotate) your head to the right, while keeping your finger exactly where it is. If you move your head far enough to the right, the object that you originally perceived to be on the left of your finger will now be on its right. Your perception of the physical, or objective, nature of your finger and the other object—their size and location within the room—has not changed, but clearly *something* has changed, because your finger and the object have "swapped places" in your visual field (see also Figure 12).

So based on these two tests, it is certainly not true that we perceive only the physical, or objective, properties of objects. Sure, we perceive those too, like in the depicted scene in Figure 10, where we perceive the right line segment to be physically larger than the left line segment. But we *also* perceive "retinal" properties like an object's angular size. It is the latter, and not the former, that is where the illusion lays in Figure 10, and in geometrical illusions generally.

But then, how do we answer the question, "Why would we *want* to perceive what's on our retina?" The answer is that we don't want to perceive what's on our retina. What we want to perceive—in addition to physical and objective properties of objects—is *where* objects in the world are in relation to us. More specifically, we want to know what direction each of the parts of a scene are in: To our right? To our left? Directly in front of us? When we see, we don't want to see—and do not in fact see—a written list of all the physical properties of the objects in the room (including ourselves), including numbers that indicate the objects' abstract spatial coordinates. No, we want to perceive the objects arrayed physically in the world, each distinguished by its unique direction from our current position. These perceptions—the ones the illusions are getting at—are not perceptions of "what's on the retina" at all. They are perceptions of the directions from us to all the objects (and object parts) in the scene. They are perceptions of where things are relative to us. And the perception of "angular size" is just the perception of the *difference* in the direction from us to each of the opposite visible ends of an object. The retina is what happens to record directions to the objects around us, but that should not confuse us into thinking that the purpose of our perceptions of angular size is to perceive what is on the retina! If our eyeballs had evolved to scramble the optic array before light reached the retina (rather than being filled with clear water that allows light to reach the retina directly), then the retina would not be able to preserve information regarding the direction of objects. But we would still have been selected to perceive the directions of the objects in a scene—because knowing where things are is a crucial skill for survival. Our brain simply would have had to reconstitute this information at a later stage of visual processing.

In Figure 10, then, we misperceive where the two vertical line segments are within our visual field. In particular, we perceive the difference in the directions (from us) to the top and bottom of the line segment on the right to be greater than that to the line segment on the left in the depicted scene, i.e., we perceive the line segment on the right to have greater angular size. We perceive the top of the line segment on the right to be in a direction from us that is higher than that for the top of the line segment on the left (i.e., the line segment on the right extends higher in our visual field than the line segment on the left). Now *these* are real illusions.

And they are not just a case of ambiguity either. The light that reaches our eyes unambiguously possesses the information that the angular sizes of the two line segments are identical in the depicted scene; our retinas even accurately record this information. Yet we perceive their angular size to differ. This kind of perceptual phenomenon is radically different than our perception of the physical sizes of the objects in the depicted scene, something which can only be inferred (as opposed to concretely known) from the light that reaches the eye and is recorded by the retina, and something that is for this reason inherently ambiguous. One can imagine a visual system finding it difficult to work out the correct physical sizes in a scene, because each line stimulus could be a nearby small object or a distant large object, or anything in between. But it is utterly mysterious why we experience these geometrical illusions, in which there is no ambiguity. We know that the correct information is "out there" at the tips of our eyeballs, and that our retinas record this information correctly. If our brains would just use this unambiguous information, they wouldn't even need to make any complex inferences to know that the two line segments have the same angular size. And yet our brains don't do this. Instead, it is as if, in the cognitive "bank" illusion, we perceived the *written letters* to be "riverbank" in the first instance and "money bank" in the second. Now *that* would be a striking "bank" illusion. Geometrical illusions are just as striking.

Why do our brains make this mistake in Figure 10? Given how easy it would be to *not* make this mistake, one suspects that our brains must be *insisting* on making this mistake, on purpose, because it believes doing so is not a mistake at all (at least not when the brain is in a natural setting). Note that one thing about angular size perceptions—and perceptions of directions to objects generally—is that they change very quickly in real life. Physical or objective properties, on the other hand, tend to stay the same. When you move around, the directions from you to all the objects in your visual field are constantly in flux, which provides a dilemma for your brain: How does it perceive the directions to all the surrounding objects in real-time—i.e., in the present—given that by the time the brain generates a perception, the present will have become the past? The answer, as we have seen, is to foresee the future. Future-seeing would be crucial for angular size perceptions because the directions to objects change so quickly, but future-seeing would not be as crucial for perceptions of the physical sizes of objects because physical sizes of objects tend to stay constant in the short term.

Blur Control

How does our brain carry out the complex future-seeing necessary to perceive the present position of each object in the scene? There are two things that make this much easier on our brains. The first is that the world has a tendency to leave breadcrumbs on our retinas, radi-

FIGURE 11. *Speed lines are commonly used in comics and cartoons to depict speed. These lines mimic the blur that occurs on the retina when objects move.*

cally simplifying future-seeing. And the second is that many of the changes we need to foresee are under *our* control. Breadcrumbs first.

Unless you're an aging actress in an infomercial, blur is usually a bad thing. We're all familiar with the disappointment of taking a picture of a moving object, only to get a blurry streak in the developed photograph. The same thing happens on the retina: it registers optic streaks (breadcrumbs) that trail the position of sufficiently fast-moving objects. The faster an object flies across the retina, the more optic blur there will be on the retina. Cartoonists have long known this, and often draw blur lines trailing moving objects to help make them appear speedier, like in Figure 11. This makes future-seeing easy; the direction and length of the streak tells the brain which direction the object is moving and at what speed.

This point is crucial to understanding geometrical illusions like the one in Figure 10 (and those in Figures 1 and 2). There's no movement in these illusions, which would at first glance make future-seeing a non-starter for explaining them. But if the drawing in Figure 11 can suggest movement without any actual movement taking place—namely, by using blur streaks—then perhaps something similar is occurring in these geometrical illusions. But before we ask how geometrical illusions might be enough like optic blur to fool the brain into believing they involve movement there is something other than being able to follow the breadcrumbs of optic blur that makes much of our future-seeing easy: we are often in control of the aspects of the world that we need to future-see.

I predict that I will wiggle my nose one second after writing the period at the end of this sentence. There, I just wiggled my nose. I'm amazing—I foresaw my nose wiggling well before it actually happened! You're not impressed? Of course not. The hard part about foreseeing the future is that it tends to be out of your control. If you confine your future-seeing attempts to cases where you're in charge, then it becomes easy. Predicting whether my own nose will wiggle after typing a period is considerably easier than predicting whether someone else's will, simply because I can *make* my prediction come true. Therefore we should be especially good at perceiving the present in cases where we are changing the world through our own actions. If changing the visual properties of a scene is under *our* control, then we should be particularly competent at foreseeing the future of those properties so as to accurately perceive the present.

What kinds of visual properties *are* under our control? Look up from this book and observe the room around you, while keeping your head still. What the objects around you are doing is mostly out of your control: the couch sits and stays, and the dog does not. The visual properties of these and other objects—their color, shape, and location in the world—are not up to you.

Now rotate your eyes to the right, while keeping your head still. Objects that were in the far right side of your visual field are now on the left. Next, slide your head from side to side. When you do so, the visual scene distorts in a characteristic way (called parallax), the nearer objects making larger sweeps across your visual field than the

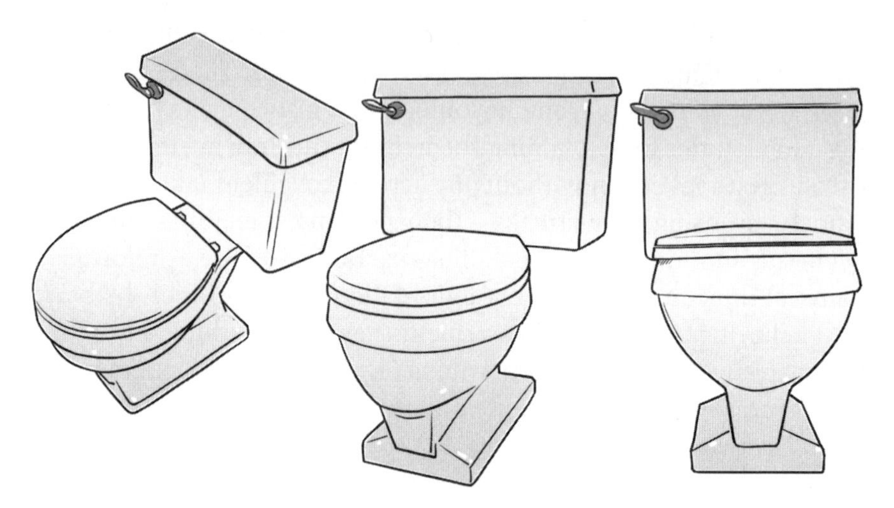

FIGURE 12. *As you move around your bathroom, you perceive the objective properties of the toilet (its width, height, color, etc.) to stay constant (historically called "constancy"), but your perception of the toilet is nevertheless changing. That is, as you move around, where each point on the toilet is relative to you changes... in a way you have control over.*

farther objects. These visual properties are ones *you* are in control of because they are not objective properties but ones concerning how the objects are located in the world in relation *to you*. You can't order a tree to change color, but you can force it to be on the left side of your visual field by facing to its right, and you can force it to fill up less of your visual field by walking farther away from it.

Now, rather than just rotating your head or moving it from side to side, get up off your seat and move around the room you're in—which for many of you may be the bathroom. Every little move your body, head, and eyes make leads to changes in your perceptions. Sure, you perceive the toilet to have certain objective properties no matter how you look at it, such as its luster, roundness, and lavender color. But exactly how it appears to you in your visual field—its location relative to you, its shape from where you stand, and how much of your visual field it fills—is constantly changing as you move about, as illustrated in Figure 12. And, importantly, *you* are the director of these changes to your visual perception. This makes it an easy matter to foresee the future for these properties, and therefore to accurately perceive the present.

Recall that the illusion in Figure 10 stems from how we perceive the locations of objects in relation to us, but not those objects' physical properties. Because the former change more quickly than the latter, it is the former that we most need to future-see. But now we realize that these visual properties are very often the easiest ones to future-see because *we* decide how they change. Lucky for us!

Foreseeing the future visual consequences of our own movements (or "self-motion") is, as already mentioned, easy. One kind of self-motion is especially easy to foresee, and disproportionately important: forward motion. The ease of prediction here arises from the fact that, as you move forward, objects flow outward in your visual field; they become larger in your visual field as you approach them and move more quickly in your visual field the closer they are to you. No matter what their objective properties are, their location in your visual field changes in a characteristic way as you move forward, and thanks to hundreds of millions of years of practice, our brains are very well adapted to this.

Forward motion is exceptionally important for two reasons: First, it is the most common motion that we, and most animals, engage in. Second, we run the risk of colliding with whatever objects we approach. Perceiving the present is crucial for quick and agile forward movement, and particularly for not getting seriously hurt. (While you can also get seriously hurt by moving backward, you don't have eyes on the back of your head, and so getting your visual perception right for backward motion, and the *inward* optic flow one experiences, won't benefit you all that much.)

I've just pointed out two aids to future-seeing, blur (or "breadcrumbs") and self-motion. Blur allows our brains to "read off" the future from the retina, and self-motion allows our brains to simply decide what the future will be. And they often occur at the same time—our self-motion creates blur we're in control of—simplifying matters even further. Recall that one of the consequences of blur was that if you add blur to a stationary image, the brain can be tricked into thinking motion is occurring. Can we build a stationary image that tricks the brain into thinking it's moving forward?

Figure 13a does just that. It is a photograph taken from a car driving forward, and certainly gives a strong impression of forward mo-

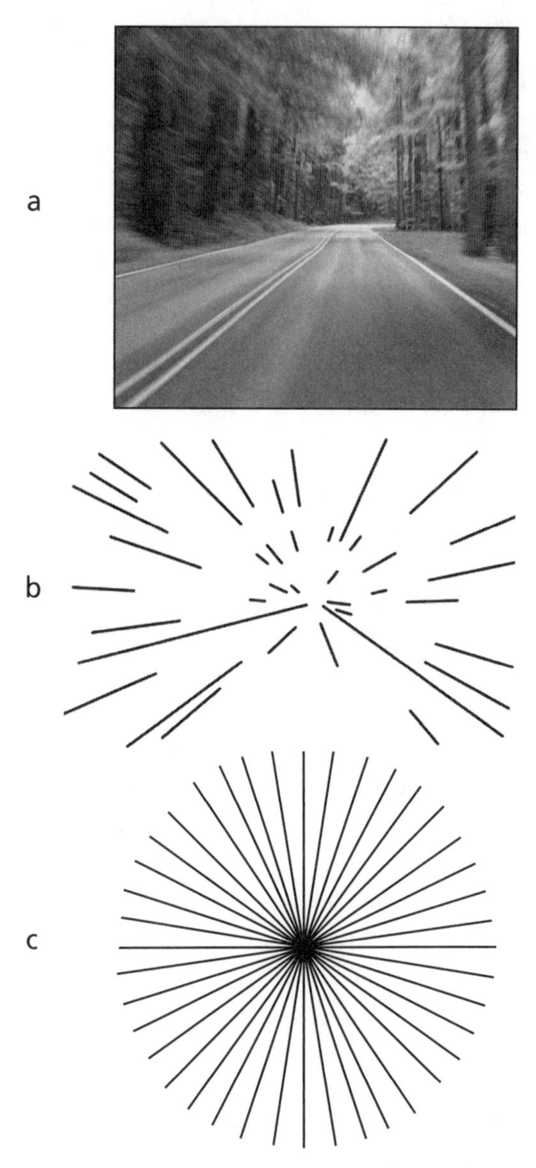

FIGURE 13. *An illustration of how lines radiating from a point (i.e., lines sharing a vanishing point) have the same structure as optic blur lines on the retina when an observer moves forward.* (**a**) *A forward-movement image, where the blur lines are directed outward, away from the direction of motion. The radial blur "lines" you see in photographs also occur on the retina.* (**b**) *The pattern of contours in the visual field during the forward movement in the image in (a).* (**c**) *An even more abstract version of (a) and (b)—a generic set of radiating lines.*

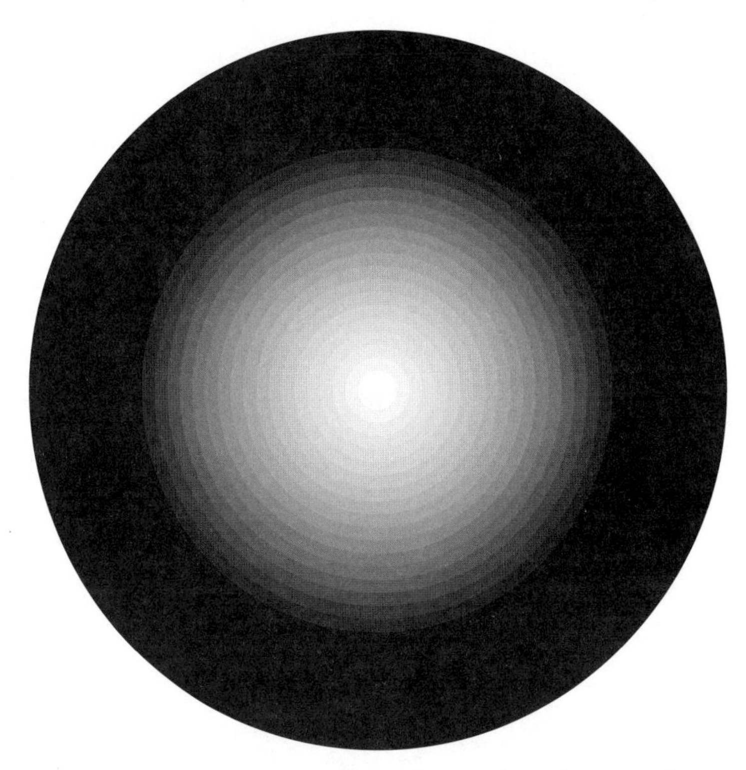

FIGURE 14. *A looming illusion invented by a student of mine, a former Duke undergraduate, David Widders. If you move your head toward the center, the brightness in the center appears to spread outward, filling the circle. If you move your head backward, it causes the dark perimeter to appear as if it is contracting toward the center.*

tion. Key to this impression is the arrangement of the optic blurs: they are arranged in a radial pattern that emanates from the center, the point toward which the car is moving. An impression of these optic streaks (with other contours, such as the sides of the road and the road's center dotted line) is shown in Figure 13b. Figure 13c is an even more abstract image of the original in Figure 13a. It shows the overall organization of the motion streaks, namely the "raw" radial line, or spoke, pattern.

Now we're getting somewhere. The classic geometrical illusions almost always possess diagonal "spoke" lines, and in light of what we have just learned, perhaps the brain is seeing the illusions in part be-

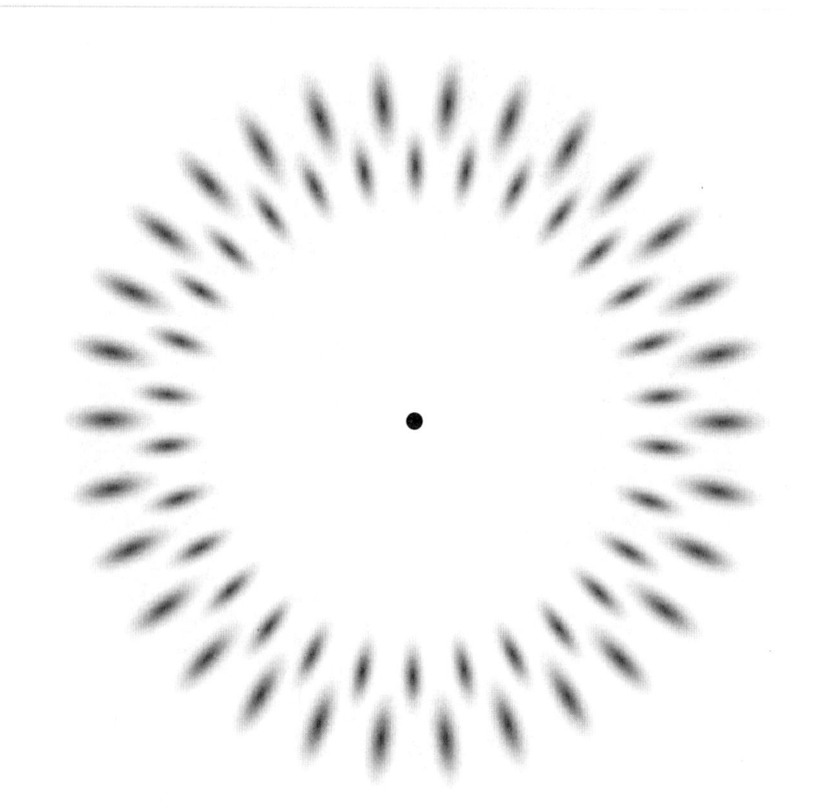

FIGURE 15. *Another illusion invented by Dave Widders. If you loom toward the center point (short, quick movement works best), the blobs flow outward faster than they should.*

cause it interprets those lines as motion streaks due to forward motion! This is one of the most critical steps in understanding classic geometrical illusions. But before we discuss these illusions in detail, we will look at a few illusions where the observer must loom his or her head forward for them to work, and show how they differ little from classic, stationary geometrical illusions.

The Looming Future

Dave Widders is a singer, guitarist, and songwriter with a growing fan base around the world. But before he was recording albums, he used his talents to study illusions with me. When he came to work with me in 2001 as an undergraduate student at Duke University, he was intent upon finding illusions induced by forward movement, and I was impressed that he quickly discovered two striking ones, the first of which I call the Dave Widders Ball Illusion. A version of it is shown in Figure 14. If you loom toward the center (that is, if you focus on the center and move your head quickly forward), the light region in the center of the ball appears to spread outward, nearly filling the ball. Interestingly, this is what *would* happen if the image were a real, three-dimensional scene we were moving through, in which the environment's grayscale changed from, in this case, black to white. But in this case the grayscale gradient is on a flat piece of paper, so it is an illusory perception rather than a true one. It also works in reverse: if you move your head quickly backward, away from the ball, the dark, outer part appears to contract toward the center, consistent with the way your environment would change within your visual field if you were moving backward in the real world.

The second illusion he found I've named the Dave Widders Fuzzy Blob Illusion, and is shown in Figure 15. If you loom toward the center in a short, fairly quick, movement, the blobs appear to flow outward faster and farther than they should—another case of perceptual overshoot. It is as if the blobs' fuzziness is misinterpreted as the blurriness that occurs in a snapshot when its subject is in motion. As a result, the short looming movement leads to the appearance that the blobs are moving at disproportionately large velocities.

In the two Widders illusions, our brain thinks we're moving forward because we actually are moving; we're looming forward. The images simply create the illusion that we are moving farther than we actually are. Might it be possible to create a static image where we perceive its contents to be moving outward even though we remain still? To make this work, the image would need to suggest that it is a "snapshot" taken during forward movement, and one very strong cue that a snapshot is taken during forward movement is optic blur. Figure 7 showed that adding blur-trails to objects leads us to perceive static images as possessing motion. Figure 16 is similar to Figure 7, except that it is consistent with forward movement, and many observers perceive the objects to be getting closer and expanding outward as if the observer were moving forward.

FIGURE 16. *Staring at the figure can sometimes lead to the perception that the objects are getting closer to you and moving outward in your visual field. This is what we would expect to happen if our brain were trying to perceive the present because the image is consistent with what we would perceive were objects moving toward us in the real world. The artist Akiyoshi Kitaoka created a similar illusion in the shape of a flower, called "Chrysanthemums."*

We can, then, successfully create a looming illusion without the actual looming! In a sense, this one-liner encapsulates a key insight for explaining classic geometrical illusions like those shown in Figures 1, 2, and 10. What I realized was that, if those radial lines *were* tricking the eye and brain into thinking that the "retinal snapshot" was taken while the observer was in forward motion, then I could explain those illusions—and a large variety of other classic geometrical illusions. And future-seeing was crucial to the hypothesis.

Consider Figure 17. At the top is a variant of a classic geometrical

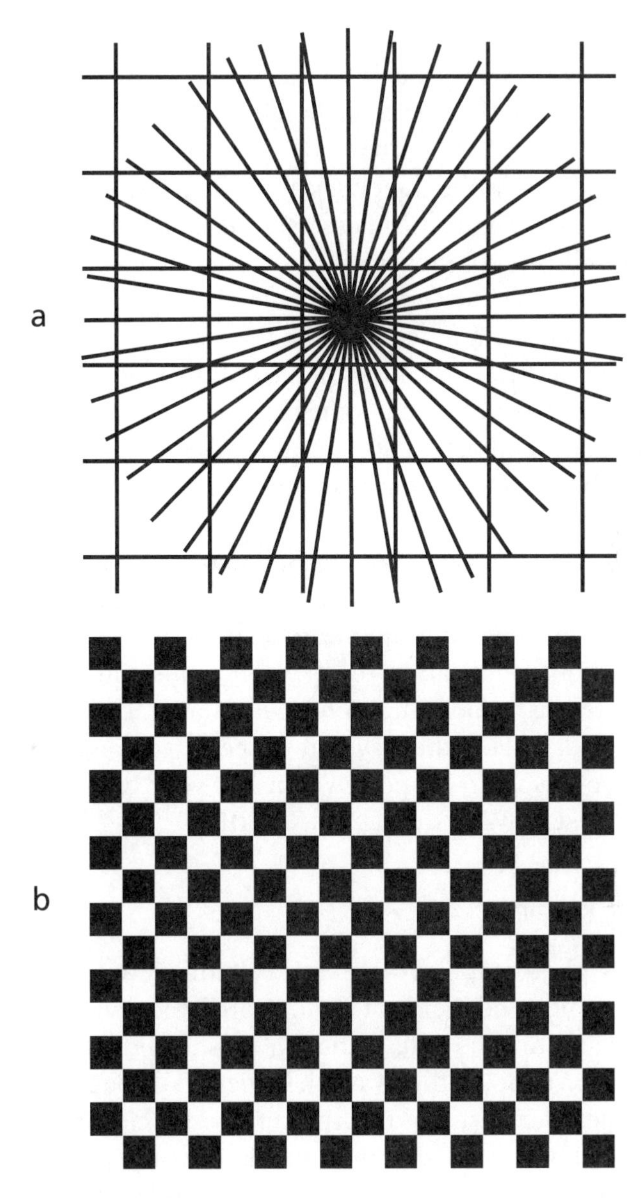

FIGURE 17. (a) *The grid appears distorted, as if the lines are bulging away from the center and toward you.* (b) *If you loom toward this grid, it appears to bulge toward you, distorting similarly to the grid in (a). This looming illusion was first noticed by Chris Foster and Eric L. Altschuler. The similarities between the two suggest that your brain is tricked into believing that the grid in (a) is moving closer to you, despite it being a static image.*

illusion in which the vertical and horizontal lines near the center of the display appear to bow away from the center. Many observers also perceive the center of the grid to be bulging slightly toward them.

Now, in light of this misperception, consider the checkerboard grid in Figure 17b. Unlike the classical geometrical illusion in Figure 17a, the grid in Figure 17b possesses no radial lines, so the image creates no illusion of movement. However, now loom toward it; it may help to fixate your eyes on a single black square. Two things are apparent: First, as was the case for the static illusion in Figure 17a, the straight vertical and horizontal contours bow away from the point where you have directed your focus. Second, and even more apparent, the checkerboard grid appears to bulge toward you—an effect noticed by vision scientists Chris Foster and Eric L. Altschuler in 2001.

The illusions created by a static rectangular grid placed on an array of radial lines (Figure 17a) is qualitatively similar to the illusion created by dynamically looming toward a checkerboard grid (Figure 17b). The obvious implication to draw from this similarity is that the visual system is responding to *both* cases as if it thinks you are looming forward. In fact, the way you misperceive these illusions is just the way you *ought* to misperceive them if your brain is trying to perceive the present. We'll discuss this shortly.

Non-Euclidean Present

Go back to the photograph of the crystal ball in Figure 1 at the beginning of this chapter, and look at it again. Notice the dividers between screens visible within the ball. Although those dividers are actually vertical, they don't appear that way when viewed through the crystal ball: they appear farthest apart at the center and closer to one another above and below. Even though the dividers are parallel to one another in reality, and even though they are parallel to one another at the center of the image within the crystal ball, this is not the case elsewhere in the image. Now look again at the geometrical illusion below the crystal ball in Figure 1. Just as the two vertical dividers don't appear to be vertical because of the crystal ball, the two vertical lines in the drawing don't appear vertical because of the illusion.

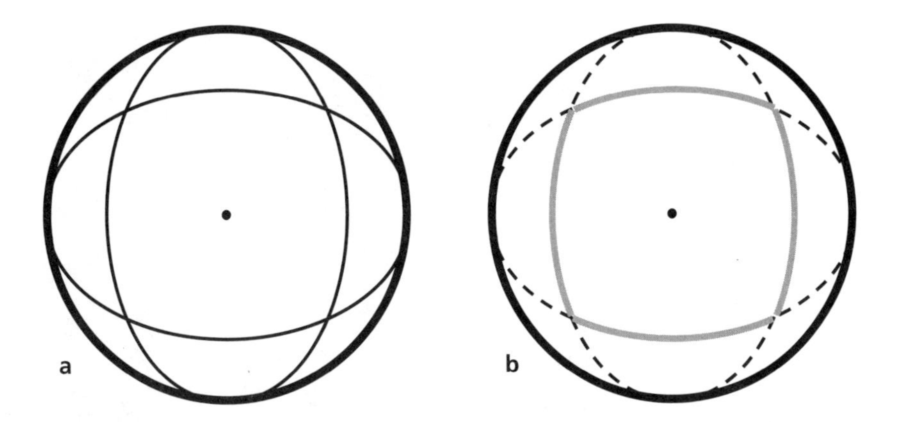

FIGURE 18. (**a**) *Two vertical lines and two horizontal lines drawn on a ball. One can see that neither pair of lines on the sphere is traditionally parallel. This is one characteristic of non-Euclidean geometry, the spherical geometry of our visual field.* (**b**) *When concentrating on the square built by the intersection of the lines in (a), one can see that each of the square's angles is larger than a right angle, another characteristic of spherical geometry.*

This is not entirely a coincidence. The crystal ball acts like a fish-eye lens, giving us a much wider and taller view than we normally possess. Inside the crystal ball, one is able to see all the objects on the other side of the ball—even those objects that are nearly overhead or are so far to the side that they appear only in our peripheral vision. The resulting image appears distorted—although, in a sense, it is not distorted at all. Allow me to explain. The two upper parts of the dividers appear to be closer to one another in the crystal ball, but this is actually true in terms of the dividers' relative distance within our visual field (i.e., the angular distance). The angular distance between the dividers is largest at eye level, but decreases as you look higher and higher. To help drive home the point, imagine that the dividers were to extend miles into the sky. As you gazed up at them, they would look like railroad tracks headed away from you, converging within your visual field. Which is just to say that the difference in the direction from you to the two dividers varies depending on how high you look: at eye level the difference in direction (from you) to each divider is high, but the difference decreases toward zero the higher you look.

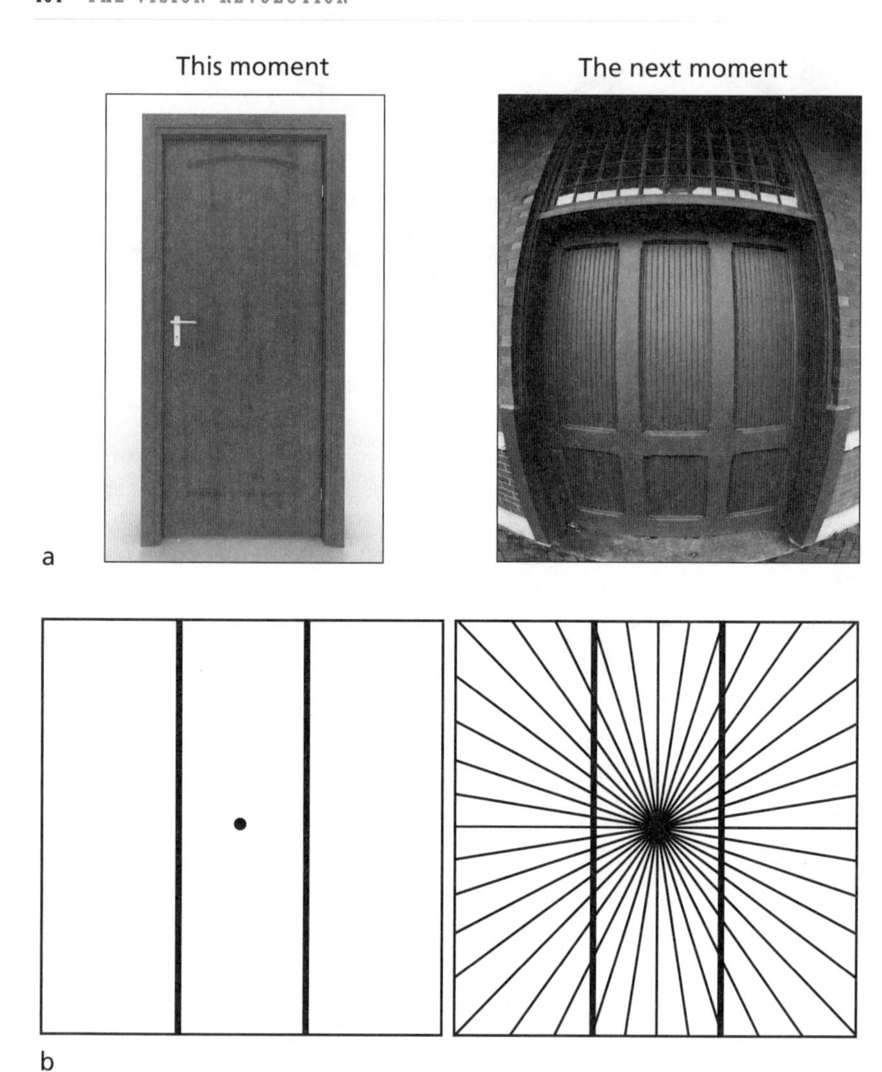

FIGURE 19. *An illustration of how the directions from you to the sides of a doorway change as you approach, from one moment to the next.* (a) *When you approach the doorway, the sides flow outward, but do so more at eye level (the middle here) than above or below. The photo on the right is of a door as seen through a fish-eye lens, which exaggerates the distortions that occur as you get closer for the sake of illustration.* (b) *On the left are the sides of a doorway as seen in the left-hand image in Figure 19a. The same two vertical lines are shown on the right, but with the addition of radial lines, which are interpreted by the brain as optic blur due to forward motion. One can see that the vertical lines seem to bow out the same way they do as you approach a doorway, as in the right-hand image in Figure 19a.*

These characteristics of your visual field are due to the geometry of spheres. The space of directions from you to all the objects around you is inherently shaped like the surface of a sphere (because the surface of a sphere encapsulates all possible directions you can point to around yourself), and thus a non-Euclidean geometry is the relevant one to use in discussing how objects change over time in your visual field. (Euclidean geometry is what you were taught as a kid, and concerns flat surfaces. Surfaces of spheres are one kind of non-Euclidean geometry.) Figure 18a shows a ball with two vertical lines drawn top to bottom and two horizontal lines drawn left to right. If you encountered the vertical lines while at the equator, you'd find that they were parallel to one another. And if you didn't realize you were standing on a sphere, you might assume that the lines would never intersect. But on a sphere they would. Figure 18b emphasizes how the corners of a square built by the intersection of the lines in Figure 18a have angles greater than ninety degrees, one of the signature features of spherical geometry.

You can see the spherical geometry of our visual field in action by observing how the direction from you to the sides of the doorway change as you walk toward it. When you are far away from the doorway, the sides of the doorway in your visual field are very nearly parallel to one another, as indicated by the left-hand image in Figure 19a. As you approach the doorway, the sides flow outward in your visual field. However, the sides of the doorway at eye level flow outward the fastest, leading continually, in the "next moment" (i.e., in about a tenth of a second), to an image like the one in the crystal ball at the start of the chapter. The right side of Figure 19a illustrates (in exaggerated form, via a fish-eye crystal-ball-like lens) how the directions from you to the door change. The difference in direction from you to the opposite sides of the door is suddenly greater at eye level than it is above or below eye level. That is, the sides of the doorway bow outward in the next moment.

Let's return now to the classical geometric illusions. Consider now the left-hand image in Figure 19b. The pair of vertical lines is analogous to the sides of the doorway in the left-hand image in Figure 19a. If we believe my idea that the radial lines of classic geometrical illusions look like the optic blur pattern that occurs on the retina during forward movement, then these lines trick the brain into thinking it is moving forward. And because the brain is attempting to perceive the present via future-seeing, it creates a perception not of the way the vertical lines actually are (namely, the left-hand image in Figure 19b), but of how they will be in the next moment. We have seen how vertical lines in the world *would* change were we moving forward; because of the spherical geometry of our visual field, they would bow outward as illustrated in the right-hand image in Figure 19a. And that is exactly what we perceive when we overlay the pair of vertical lines with radial lines, as shown in the right-hand image in Figure 19b.

That is how we can explain classic geometrical illusions like those in Figures 1b, 2, 10, and 17a. And all we needed was an understanding of optic blur, future-seeing, and non-Euclidean spherical geometry!

Who would have thought—crystal balls have something to do with future-seeing after all! When we move forward, the directions from you to all the parts of the scene change, and the manner in which these things change is governed by spherical geometry. It can be difficult to consciously appreciate these changes because we typically look at only the small piece of the world that lies directly in front of us at any one time. If, however, we look at the world through a crystal ball, the non-Euclidean distortions immediately become apparent. And if we move forward while holding our crystal balls in front of us, they provide an advanced image of the way the world will distort in the moments to come. In this sense, crystal balls really *can* be used to see the future!

Illusions United

Future-seeing and the way it allows us to perceive the present are interesting in and of themselves. But they are also important because they carry implications for our understanding of many mysterious visual effects, especially the enigmatic classic geometrical illusions. In fact, these geometrical illusions were what originally prompted my perceiving-the-present hypothesis. I realized that these illusions could be explained as the visual system trying to perceive the present while being tricked (by radial lines) into thinking it was moving forward. What I had not anticipated was the extent to which this idea also helped explain other classes of illusions. A couple years after coming to understand geometrical illusions as mistaken perceptions of the present, I began to notice that other illusions, many seemingly quite different, possessed certain similarities to the classic geometrical illusions I was studying, and I realized that this same idea could act as a "super" prediction for illusions, one that would indicate the pattern of illusions that researchers should have found across many different kinds of stimuli. That is, I realized that my theory about classic geometrical illusions was really the beginning of what might be called a "grand unified theory" of illusions. Now, if anyone begins telling you that they have a grand unified theory of something, my usual advice is to nod approvingly, ask to use the restroom, and then sneak out the window. But if crystal balls, future-seeing, *Men in Black*–style mental video tape editing, head-looming, and non-Euclidean geometry haven't frightened you away, you're probably here for the duration.

The key idea is this: Classical geometrical illusions occur because (1) the radial blur cues the visual system as to the direction in which the observer is headed, and (2) where objects appear in the visual field changes in a predictable fashion when an observer moves forward. But radial blur is just one of many optical properties that occur when moving forward and that cue the observer's visual system about the probable direction of motion. And where an object is located in the visual field is just one of many optical properties that change in a predictable fashion as the object or the observer moves. If we take any of our direction-of-motion cues and pair it with any of the optical-properties-that-predictably-change, then we should get more illusions. But this theory would only work if the reason classical geometrical illusions work is because our brains are perceiving the present by foreseeing the future; otherwise, we should not find illusions from all of these combinations. And, as it turns out, we do.

The row headings in Figure 20 list seven direction-of-motion cues. The sixth row contains the radial spokes (optic blur) relevant to classic geometrical illusions, but now we have six other potential direction-of-motion cues. For example, the first row, "angular size" (how much of your visual field is filled by an object), is on the list of direction-of-motion cues because moving forward causes the angular size of an object to increase as you approach it. Objects with smaller angular sizes tend to be nearer to the direction you are headed in than objects with larger angular sizes. You can see this in Figure 13a. The other row headings are similar in that, in each case, they provide cues as to what direction the observer might be going. In short, in comparison to other objects, objects in the direction you are headed tend to have smaller angular sizes, have lower angular speeds (how fast objects move across your visual field), have greater luminance contrasts (because of lower speeds and less blurring), have greater distances from the observer, have lower eccentricity (i.e., people tend to look where they're going), be closer to the vanishing point of optic blur lines, and be closer to the point from which real live dynamic optic flow (the pattern of objects' apparent motion) emanates. (Luminance contrast, or just contrast, is the brightness difference from the background.)

The column headings in Figure 20 list four optical-properties-that-predictably-change. The first is angular size, and is the property relevant to the classic geometrical illusions, but now we have three others. Angular speeds in the visual field are predictable when an observer is moving forward—they speed up—but the acceleration slows so that while they are at their fastest by the time they pass, they are no longer increasing in speed. A consequence is that if there are two objects that are similarly distant from you, the one nearer to where you are headed will accelerate more in the next moment. It will also have greater angular size (the first column heading), have lower contrast (because objects moving faster become more blurred), and appear to become closer more quickly in comparison to the object that you are not headed as directly toward.

In each of the boxes in the table in Figure 20, there are two "target objects" that are similar in regard to the property in the column heading. In the first column the two objects always have the same angular size. In the second column, there are always two identical arrows, indicating that the two objects are moving with the same angular velocity. (Any arrows shown in these figures indicate that the stimulus would be a movie rather than a static image, with objects

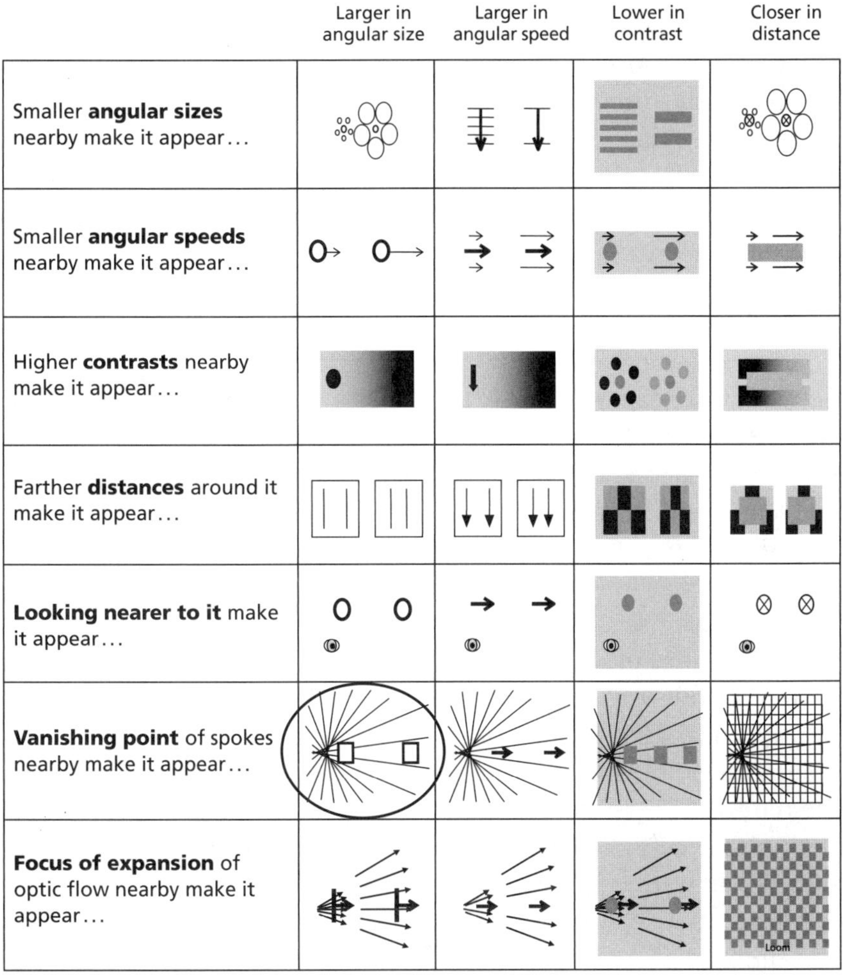

	Larger in angular size	Larger in angular speed	Lower in contrast	Closer in distance
Smaller **angular sizes** nearby make it appear...				
Smaller **angular speeds** nearby make it appear...				
Higher **contrasts** nearby make it appear...				
Farther **distances** around it make it appear...				
Looking nearer to it make it appear...				
Vanishing point of spokes nearby make it appear...				
Focus of expansion of optic flow nearby make it appear...				Loom

FIGURE 20. *An illustration depicting the unifying power of perceiving-the-present across the butterfly collection of illusions gathered over the history of vision science. There are the same icons that appear in Figure 9, but they are now systematically organized in light of the idea that our visual systems attempt to perceive the present via future-seeing, particularly in circumstances where the observer is moving forward. Each icon represents a class of illusions. For example, the first icon in the first row is for the class of illusions where angular size variations in a drawing lead to the perception that two objects of the same angular size differ in angular size. The one below that stands for the class of illusions where speed variations of objects in a movie lead to the perception that two objects of the same angular size differ in angular size. The classic geometrical illusions, circled above, turn out to be just one of twenty-eight classes of illusions explained by our brains' efforts at perceiving the present.*

moving in the direction indicated with a speed proportional to the arrow length.) Also, for each box in the table, the figure shown is set up so that the direction-of-motion cues (indicated by the row label) suggests that the observer is heading more toward the target on the left. For each of these stimuli, then, the prediction is that the target object on the left is closer to the observer's direction of motion, and will therefore, in the next moment, appear to be larger (column 1), be faster (column 2), have less contrast (column 3), and be closer (column 4).

Seven rows and four columns adds up to twenty-eight predictions. So we can see that the classic geometrical illusions are just a special case (shown circled in Figure 20) of what turns out to be a radically more general theory. Or, rather than thinking about it as twenty-eight different predictions, each for a specific kind of illusion, we can think about it as a single "large" prediction about the *pattern* of illusions.

To my surprise, after an intense review of the illusions discovered over the course of vision science's history, I found that the predicted pattern of illusions exists. Recall that the hypothesis was originally developed with only the classic geometrical illusions in mind. This makes the success of this prediction even more powerful as support for the hypothesis, because it means I couldn't have rigged my hypothesis—I couldn't have come up with a hypothesis to fit the data rather than testing to see if the data fit my hypothesis. The success of this prediction is also of interest because it allows us to systematize and unify so many illusions that heretofore seemed unrelated.

Past, Future, and Present

The story of this chapter has been about the need to perceive the present so that we might control it. Earlier we saw that in order to perceive the present you need to see the future, and we have since discussed how one of the simplest ways to see the future is to *make* the future happen—something that in many cases is actually quite easy, in particular when you move forward. He who controls the future, then, controls the present. And, although it hasn't been the primary issue discussed in this chapter, seeing the future is only possible because our visual system appreciates how visual features change over time—something that must have been learned from past experiences, either our personal past experiences or the past experiences accumulated by our ancestors and instilled in our genes. He who controls the past, then, controls the future. This leaves us with a twist on Orwell's "He who controls the present controls the past. He who controls the past controls the future." Instead, we have, "He who controls the past controls the future. He who controls the future controls the present." And at the end of the day, isn't it really the present we most want to control?

CHAPTER 4
Spirit-Reading

"Ssh!" said her Daddy, and frowned to himself, but Taffy was too excited to notice.

"That's quite easy," she said, scratching on the bark.

"Eh, what?" said her Daddy. "I meant I was thinking, and didn't want to be disturbed."

"It's a noise just the same. It's the noise a snake makes, Daddy, when it is thinking and doesn't want to be disturbed. Let's make the ssh-noise a snake. Will this do?" And she drew this: "S."

—Rudyard Kipling, "How the Alphabet Was Made" (*Just So Stories*)

Super Reading Medium

Communicating with the dead is a standard job requirement for a psychic. The infamous medium John Edward, of the television show *Crossing Over*, claims to hear what the deceased family members of his studio audience have to say, but generally, hearing the thoughts of the dead is one superpower it appears we do *not* possess. Surely this superpower remains firmly in the realm of fiction (Edward included). However, a little thought reveals that we hear the thoughts of the dead all the time... simply by reading. With the invention of writing, the dead were suddenly able to speak to the living. (Progress in communicating in the other direction has been slower going.) For all you know, *I'm* dead, and you're exercising your spirit-reading skills right now. Good for you!

Before the advent of writing, to have our thoughts live on after our deaths we had to invent a story or tune catchy enough that people would recite it around the fire for generations. Only a few were lucky enough to create a song or story with such legs (e.g., Homer's *Iliad*). And if our ancestors were anything like us, their greatest hits probably included "ooh-la-la" and "my baby left me" much more often than "here's my unsolicited advice" and "beware of milk-colored berries." Using your children as audio tape in this fashion is probably futile (and aren't they just as likely to purposely say the opposite?), but at least it relies on spoken words, something readily understandable by future generations. The problem is getting your voice to last. Voices are too light and insubstantial; trying to get them to last is like a quarterback trying to throw a seventy-yard pass with a marshmallow. Marshmallows are easy enough to hold, but impossible to throw far. Perhaps, if you yelled loudly enough during a heavy volcanic ash storm, the rapidly accumulating layers of ash would preserve the ripples from the sound waves your voice would make, and a clever archeologist decoder might one day recover them. Unfortunately, much of what you're likely to say in such circumstances would be unrepeatable in polite company.

What prehistoric people *did* successfully leave behind tended to be solid and sturdy, like Stonehenge or the moai statues of Easter Island. Passes like this get all the way to the end zone, except now the quarterback is throwing an eighteen-wheeler filled with two tons of marshmallows, and no one on the team is able to catch it. Massive monuments are great if your goal is to impress the neighboring tribes or to brag about how powerful you are to posterity. But if your goal is to have your message actually be understood, this tactic is worse than writing abstruse poetry, and literally much heavier. The only thing we're sure of from such communication is that those prehistoric people had too much free time on their hands. Not the most informative spirit-reading.

The invention of writing changed spirit-reading forever. It also changed the world. Reading now pervades every aspect of our daily lives; one would be hard-pressed to find a room in a modern house without words written somewhere inside, probably lots of them. Many of us now read more sentences in a day than we listen to. And the act of reading is a complicated one. It involves processing thousands of tiny shapes in a short period of time. In this book alone there are more than 300,000 strokes, and many long novels include well over one million. And not only are we human beings highly competent readers, but our brains appear to have regions devoted specifically to recognizing words. Considering all this, a Martian just beginning to study us humans might be excused for concluding that we had evolved to read. But of course, we haven't. Reading and writing are recent human inventions, developing at most only a few thousand years ago, and much more recently in many parts of the world. We are reading using the eyes and brains of our illiterate ancestors. This brings us to a deep mystery: Why are we so good at such an unnatural act? We read as if we were designed to do so, but we have *not* been designed to do so. How did we come to have this superpower?

Reading as a superpower? Isn't this, you might ask, a bit of an exaggeration? Not at all! To better appreciate it, next time you have your illiterate caveman neighbors over to the house—the ones who always bring the delicious cave-made Bundt cake—wow them with how quickly you can transmit information between you and your spouse without speaking to one another...by writing and reading. They'll certainly be impressed—but not by your use of symbols. They leave symbols for one another all the time, like a shrunken head in front of the cave to indicate the other is at the witchdoctor's. They have spoken language, after all, and realize that the sounds they utter, too, are symbols. What will amaze them about your parlor trick is how freakishly efficient you are at it. How did your spouse read out the words from the page so fast? Although they appreciate that there's nothing spooky in principle about leaving marks on a page for someone to interpret later, they conclude that you are much too good at it, and, despite your protestations, you must be magical shamans of some kind. They also don't fail to notice that your special power would work even if the writer was far away. Or long dead. Their hairs stand on end, the conversation becomes forced, and they skip dessert. Later you notice that their cavekids don't come around to throw spears at your kids any more. As the saying goes, one generation's maelstrom is a later generation's hot tub. We're too experienced with writing to appreciate just how super it is; not so, your caveman neighbors.

We have the superpower of reading not because we evolved to read—and certainly not because we're magical—but because writing evolved to work well with the eye. Just as some galaxy locals interpreted Captain Kirk's technology as magic, your neighbors are falsely giving you credit for the power to communicate when the real credit should go to the technology of writing. Written language is not some untested technology, but one that has been honed over many centuries, even millennia, by cultural evolution. Writing systems and visual signs tended to change over time, the better variants surviving while the worse ones did not. The result is written language that allows meanings to flow almost effortlessly off the page and straight into our minds. Instead of seeing a morass of squiggles we see the thoughts of the writer, almost as if he or she is whispering directly into our ears.

As we'll see in this chapter, the special trick behind the technology is that human visual signs have evolved to look like nature. Why? Because *nature* is what we have evolved over millions of years to be good at seeing. We are amazingly good at reading the words of the dead (and, of course, the living) not because we evolved to be spirit-readers, but because we evolved for millions of years to be good at visually processing nature. Writing has just evolved to tap into this ability. Our power to quickly process thousands of tiny shapes on paper is our greatest superpower of all, changing our lives more than the other superpowers. Literacy is power, and it's power we have all because our eyes evolved to see the natural shapes around us and we, in turn, put those shapes to paper.

Good Listening

"How did your date go?" I asked.

"Great. Wow. What a guy!" she replied. "He listened so attentively the entire dinner, just nodding and never interrupting, and—"

"Never interrupting?" I interjected.

"That's right. So supportive and interested. And so in tune with me, always getting me without even needing to ask me questions, and his—"

"He asked you no questions?" I interrupted, both eyebrows now raised.

"Yes! *That's* how close the emotional connection was!"

It struck me that any emotional connection she felt was a misreading of his eyes glazing over, because her date was clearly *not* listening. At least not to her! I didn't mention to her that the big game had been on during her dinner, and I wondered whether her date might have been wearing a tiny earphone.

Good listeners don't just sit back and listen. Instead, they are dynamically engaged in the conversation. *I'm* a good listener in the fictional conversation above. I'm interrupting, but in ways that show I'm hearing what she's saying. My questions are allowing me to get more details of the story where I need them, in this case about her date's conversational style. That's what good listeners do. They rewind the story if needed, or forward it to parts they haven't heard, or ask for greater detail. Good communicators tend to be those who are capable of interacting while they are talking. If you bulldoze past all of your listener's attempts to interject, he probably won't be listening for long. Perhaps he was trying to tell you he'd heard that part before, and when you went on anyway, he tuned out. Or perhaps he was confused by something you said fifteen minutes earlier, and when he wasn't able to have you clarify, he gave up trying to make sense of what you were saying. Good listeners require good communicators. My fictional friend above appears to be a good communicator because she dynamically reacts to my queries midstream. The problem lay in her date, not her.

Even though we evolved to speak and listen, but not to read, there is a sense in which writing has allowed us to become much better listeners than speech ever did. That's because readers can easily interact with the writer, no matter how far removed the writer may be in space or time. Readers can pause the communication, skim ahead, rewind back to something not understood, and delve deeper into certain parts. We listeners can, when reading, manipulate the speaker's stream of communication far beyond what the speaker would let us get away with in conversation—"Sorry, can you repeat the part that started with, 'The first day of my trip around the world began without incident'?"—making us super-listeners and making the writer a super-communicator.

We don't always prefer reading to listening. For example, we listen to books on tape, lectures, and talk radio, and in each case the speakers are difficult to interrupt. However, even these examples help illustrate our preference for reading. Although people do sometimes listen to books on tape, they tend to use them only when reading is not possible, like when driving. When our eyes are free, we prefer to read stories rather than hear them on tape, and the market for books on tape is miniscule compared to that for hard copy books. We humans have brains that evolved to comprehend speech, and yet we prefer to listen with our eyes, even though our eyes weren't designed for this! Television and movies have an audio stream that is not easily interruptible, and we *do* like them, but that may be in part because the visual modality helps keep our attention. And although students have been listening for centuries to their professors speak, until recently with relatively little visual assistance, anyone who has sat through years of these lectures knows how often one's mind wanders and how often one is not actually listening. Talk radio is somewhat popular, and tends to be more engaging than traditional lectures, but notice that such shows go to great lengths to be conversational, using a conversational format with guests, typically encouraging conversations with callers, and often having more than one host (or a sidekick), to elicit the helpful interruptions found in good listening.

Canned speech, then, tends to be difficult to listen to, whereas genuine, dynamic, interactive conversation enables good listening. There is, however, one kind of audio stream our brains can't get enough of, where interruption is not needed for good listening and where we're quite happy not seeing anything: music. Audio tapes that give up on communication and aim only for aesthetics suddenly become easy listening. The rarity of books on tape, and the difficulty in listening to canned speech more generally, is not due to some intrinsic difficulty with hearing per se. The problem is that speech requires literal comprehension—music doesn't—and comprehension can occur most easily when the listener is able to grab the information by the scruff of the neck and manipulate it as needed to fit it into his head. Good conversation with the speaker can go a long way toward this, but you can achieve this even more efficiently by reading because reading lets you literally pick up the information with your hands and interact with it to your heart's content.

Working Hands

Having a conversation is not like passing notes in class. Although in each case two people are communicating back and forth, when passing notes you tend to do little reading and lots of wiggling—either wiggling your hand as you write a note, or twiddling your thumbs as you wait for your friend to write his. Note writing takes time, so much time that writing dominates most of the time spent passing notes back and forth, and is interspersed with only short bouts of reading. All the work is in the writing, not the reading. Conversation—i.e., people speaking to one another—is totally different, because you are either speaking or listening the whole time. Speaking flows from us effortlessly, and comes out nearly at the speed of our internal thoughts. That is, whereas writing is much more difficult and time-consuming than reading, speaking is about as easy as listening.

The reason for this has to do with the number of people we're communicating with. When we speak there are typically only a few people listening, and most often there's only one person listening (often less than that when I speak in my household). For this reason, spoken language has evolved to be a compromise between the mouth and ear: somewhat easy for the speaker to utter, and somewhat easy for the listener to comprehend. In contrast, a single *writer* can have an infinite number of readers, or "visual listeners." If writing has similarly evolved to minimize the overall efforts of the community, then the *readers'* efforts will be what has driven the evolution of writing because they are the majority. That's why, as amazing as writing may be, it is a gift more to the eye than to the hand. For example, this book took me about six months to write (and much of it was actually initially written by hand), but it may take you only six hours to read. That's an effective compromise because there will (hopefully) be many readers of this book—many more "listeners" than I would have reached had I used that six months to have individual conversations.

Is writing really for the eye, at the expense of the hands? One of the strongest arguments that this is the case appears later in this chapter, when we see that writing has been culturally selected to look like nature. That's a good thing for the eye because the eye has evolved to see nature; the hand, however, has not evolved to draw nature. Not only does writing tend to look like nature, but I have found that even visual symbols like trademark logos—which are typically never written by hand, and are selected to be easy on the eyes—use the fundamental structural shapes found in nature. And note that for some decades now much of human writing has been typed with a keyboard, not written out by hand. If the structures of letters were designed for the hand, we might expect to find, now that our hands tend to be out of the picture, that the structures of letters had changed somewhat. However, although there are now hundreds of different fonts available on computers, the fundamental structural shapes of letters have stayed the same. Shorthands, however, have been explicitly designed for the hand at the expense of the eye—they are for *recording* information quickly, not comprehending information quickly—and shorthands look radically *different* from normal writing; I have shown that they have shapes that are *not* like nature. I have also taken data from children's scribbles and shown how the fundamental structures occurring in scribbles are unlike those found in writing and in nature. Finally, one can estimate how easy a letter is to write by the number of distinct hand movements required to produce it (this counts movements resulting in strokes on the page, and also movements of the hand between touching the paper), and such estimates of "motor ease" do not help explain the kinds of shapes we find in writing. Writing is clearly not shaped for the hand.

Could culture really have given *no* thought whatsoever to the tribulations of the hand? Although selection would have favored the eye, it clearly would have done the eye no good were writing so difficult that no hand was willing to make the effort. Surely the hand must have been thrown a bone, and it probably was. The strokes in the letters you're reading, and in line drawings more generally, are quite a bit like the "contours" in the real world that occur whenever one surface stops and another starts—the edge between two walls, or the edge of your table—in that they are thin, but usually, with contours, there is no line or stroke at all (although sometimes there can be), just a sudden change in the nature of the color or texture from one region to the next. Our visual system would therefore probably prefer to look at contours, not strokes. But strokes are still fairly easy to see by the visual system, and are *much* easier for the hand to produce. After all, to draw true contours rather than strokes would require drawing one color or texture on one side of the border of the intended contour and another color or texture on the other side. I just tried to use my pen to create a vertical contour by coloring lightly to the right and more darkly to the left, but after a dozen tries I've given up. I won't even bother attempting this for an "S"! It's just too hard—which is why when we try to draw realistic scenes we often start with lines in place of contours, and only later add color and texture between the lines. That's probably why writing, too, tends to use strokes. That we use strokes and not true contours is for the benefit of the hand, but the shapes of our symbols are for the benefit of the eye, as we'll see later.

Harness the Wild Eye

You'd be surprised to see a rhinoceros with a rider on its back. In fact, a rider would seem outlandish on most large animals, whether giraffe, bison, wildebeest, bear, lion, or gorilla. But a rider on a horse, on the other hand, seems natural. Unless you grew up on a farm and regularly saw horses in the meadows, most of your experiences with horses were likely from books, television, and film, where horses typically have riders. Because of your "city folk" experiences with horses, a horse without a rider can seem downright unnatural! In fact, if aliens were observing the relationship between humans and horses back when horses were our main mode of transportation, they may have falsely concluded that horses were designed to carry humans on their backs. Of course, horses aren't born with bridles and saddles attached, and they didn't evolve to be ridden. They evolved over tens of millions of years in savannas and prairies, and it was only recently that some primates had the crazy idea to get on one. So how is it that horses became so well adapted as "automobiles" in the human world?

Horses didn't simply get pulled out of nature and plugged into society. Instead, culture had to evolve to wrap around horses, making the fit a good one. Horses had to be sired, raised, fed, housed, steered, and scooped up after. Humans had to invent countless artifacts to deal with the new tasks required to accommodate the entry of horses into society, and entire markets emerged for selling them. Diverse riding techniques were developed and taught, each having certain advantages for controlling the beasts. The shapes of our homes and cities themselves had to change: water troughs added in front of every saloon, stables stationed throughout towns and cities, streets built wide enough for carriages, and so on. That horses appear designed for riders is an illusion that arose from culture's adaptation of *society* to *horses*.

Just as horses didn't evolve to be ridden, eyes didn't evolve for the written. Your eyes reading this book are wild eyes, the same eyes and visual systems possessed by our ancient preliterate ancestors. Yet despite being born without a "bridle," your visual system is now saddled with reading. Our ability to read presents the same kind of mystery as bridled horses: How do our ancient visual systems fit so well in modern, reading-intensive society?

Eyes may seem like the natural choice for obtaining information, and indeed vision probably has inherent superiorities over touch or taste, just as horses are inherently better rides than rhinos. But just as horses don't fit efficiently into culture without culture changing to fit horses, the visual system couldn't be harnessed for reading unless culture developed writing to fit the requirements of the visual system. We didn't evolve to read, but culture has gone out of its way to create the illusion that we did. We turn next to the question of what exactly cultural evolution has done to help our visual systems read so well.

From the Hands of Babes

You might presume that a two-and-a-half-year-old girl couldn't have much to say. If I were struck on the head and reduced to infant-level intelligence for two and a half years, I'm fairly sure I wouldn't have a flood of stories to recount. None, at least, that were not considerably degrading. But there my daughter was, talking up a storm. A little about the few things that have happened to her, but mostly about things that never have, and never will: princesses, dragons, SpongeBob, stegosauruses. She's five now and there's been no letup. She's talking to me as I write this!

I just gave her a piece of paper and crayons, and although she's just begun trying her hand at writing—"cat," "dug" (dog), "saac" (snake), "flar" (flower)—she's been putting her thoughts and words to the page for a long time now—by drawing. Children invent their own writing through pictures, and because of that offer us the chance to better understand the invention of writing. Through the work of Rhoda Kellogg in the mid-twentieth century, we know that children worldwide draw very similar shapes and follow a similar developmental schedule. Since they almost certainly have not evolved to draw, these similarities are, in a sense, parallel discoveries about how to effectively communicate on paper—on how to write. Sir Herbert E. Read, an early twentieth-century professor of literature and arts, encountered Rhoda Kellogg's work late in his life, and wrote the following:

> It has been shown by several investigators, but most effectively by Mrs. Rhoda Kellogg of San Francisco, that the expressive gestures of the infant, from the moment that they can be recorded by a crayon or pencil, evolve from certain basic scribbles towards consistent symbols. Over several

> years of development such basic patterns gradually become the conscious representation of objects perceived: the substitutive sign becomes a visual image. . . . According to this hypothesis every child, in its discovery of a mode of symbolization, follows the same graphic evolution. . . . I merely want you to observe that it is universal and is found not only in the scribblings of children but everywhere the making of signs has had a symbolizing purpose—which is from the Neolithic age onwards.[1]

But aren't children's drawings just that, drawings? It's certainly true that sometimes children are just trying to depict what they see. Those are "mere" drawings. But often their drawings are aimed at *saying* something—at telling a story. When my daughter brings me her latest drawing, she usually doesn't brag about how real it looks (nor does she tell me about its composition and balance). Sure, sometimes she asks me to count how many legs her spider has, but usually I get a story. A long story. For example, here is a CliffsNotes® version of the story behind her drawing in Figure 1: In it is a house with arms and eyes; the windows have faces; it is a magic house; there is a girl holding a plate of cream puffs; two people are playing with toys at the table but a tomato exploded all over the toy; there are butterflies in the house. The drawing is intended to communicate a story, and *that* sounds an awful lot like writing.

But if my daughter were truly writing, then she'd have to be using *symbols*. Is it really plausible that small children are putting symbols on the page before they learn formal writing, as Rhoda Kellogg and Herbert Read believe? I think so. Consider that most children's drawings bear only the vaguest resemblance to the objects they are intended to depict. Look at nearly any of the objects in my daughter's drawing in Figure 1. An attempt at realism? Hardly. And we find similar kinds of symbols in cartoons drawn by adults—adults who *could* draw realistically if they wished. These cartoon symbols, like those in the first row of Figure 2a, are ridiculously poor renderings of objects. There are similar visual signs in every culture. And although you'll probably have no trouble figuring out what animals the draw-

[1] From Herbert Read, "Presidential Address to the Fourth General Assembly of the International Society for Educational Society for Education through Art." Montreal: Aug 19, 1963.

FIGURE 1. *A drawing by my daughter at five years old.*

ings are intended to symbolize, your dog would have no idea what they (or my daughter's drawings) are supposed to be. Symbols get their meaning through convention more than through resemblance. We're so used to these conventions that we are under the illusion that they actually look like the animals they refer to. But other cultures often have somewhat different conventions for their animals. For example, I find it difficult to tell what kind of animal I'm looking at in many of today's Japanese children's cartoons (you can see a few examples in the second row of Figure 2a).

The same is true for sound. We in the United States say "ribbit" to imitate a frog's call, and after growing up thinking of that as the sound frogs make, it can be hard to appreciate that frogs don't really sound at all like that. In fact, people from different cultures use different sounds to refer to frog calls, and each person is initially convinced that *their* sound is the one that most resembles a frog's. Algerians say "gar gar," Chinese say "guo guo," the English say "croak," the French say "coa-coa," Koreans say "gae-gool-gae-gool," Argentineans say "berp," Turks say "vrak vrak," and so on. The sound "ribbit" is a *symbol* for the call of the frog, not a real attempt to mimic it—just like children's drawings are symbolic rather than actually resembling the real-world objects they depict.

So children's drawings are attempts to communicate stories using symbols. That sure sounds like writing to me—at least the barest beginnings. If these little whippersnappers are smart enough to spontaneously invent writing largely on their own, perhaps it wouldn't hurt to look into the *kinds* of symbols they choose for their writing. And the kinds of symbols children choose is so obvious that it's hard to notice: Children draw *object-like* symbols for the objects in their writing. Their drawings may not look much like the objects they stand for, but they still look like *objects*, not like fractal patterns, or footprints, or scribbles, or textures. The same is true for the cartoons drawn by adults, as in Figure 2a. And we find the same so-obvious-it's-hard-to-notice phenomenon in animal calls: although there are lots of different sounds used for frog calls, they are all animal-call-*like*. All those frog calls are sounds *some* kind of animal could have made. What does this mean for writing?

Word and Object

Is drawing object-like symbols for objects mere child's play? Apparently not, because you don't find object-like symbols for objects just in kids' drawings and cartoons, but among human visual signs generally. Most non-linguistic visual signs throughout history have been object-like, such as those found in pottery, body art, religion, politics, folklore, medicine, music, architecture, trademarks, and traffic signs (see Figure 2b for a small variety). Computer desktop icons are not just object-like in appearance, they can even be moved around like objects. (By the way, *that's* why computer desktops are such useful fictions for representing a computer's innards, something we discussed in previous chapters.) Much of formal writing itself has historically been composed of objects that stand in for words, including Egyptian hieroglyphs, Sumerian cuneiform, Chinese characters, and Mesoamerican writing. Modern Chinese is still written this way, and Chinese is used by nearly half the world. In all these

a. "Language-like" art

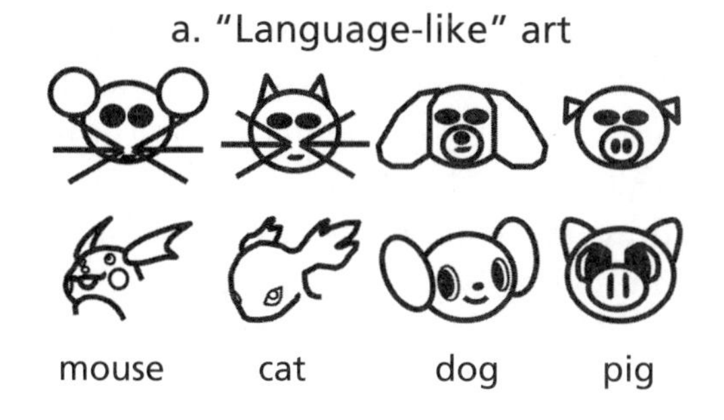

b. Non-linguistic signs

c. Logographic signs

Chinese

朋音妹爱金

Linear B

FIGURE 2. (a) *Some simple drawings of animals. Nearly anyone in the Western world would understand the drawings in the first row. The second row shows samples of the same four animals in Japanese cartoons. Each has at best minimal resemblance to the animals to which they refer.* (b) *A small sample of the variety of shapes found in human non-linguistic visual signs. Do a Web search on pictograms or symbols and you'll find tens of thousands more. Notice how these look roughly like objects.* (c) *Examples of logographic signs—signs where the symbol stands for a spoken word. As in (a) and (b), the symbols are object-like.*

writing systems we find low-complexity drawings that are used as symbols to refer to objects, and to adjectives, adverbs, verbs, and so on (see Figure 2c for several examples). Object-like symbols for objects—that trick's not just for kids.

Is there something particularly effective about drawing objects to stand for words in writing? I suspect so, and I suspect that the reason is the same as why animal-call symbols tend to be animal-call-like: we probably possess innate circuitry that responds specifically and efficiently to animal-call-like sounds, and so our brain is better able to process a spoken word that means an animal call if the word itself sounds animal-call-like. Similarly, we possess a visual system designed to recognize objects and efficiently react to the information. If a word's meaning corresponds to an object, even an abstract object, then our visual system will be better able to process and react to the written symbol for that word if the written symbol is itself object-like. Figure 3b shows a fictional case of writing using object-like symbols for words (and single strokes for "function words" like "the" and "in"). To grasp why this strategy might be a good one, consider two alternative strategies.

First, rather than drawing objects for words, we could be lazy and just draw a single line for each spoken word. The writing for "The rain in Spain stays mainly in the plain" would then look something like what is shown in Figure 3a. Shorthand, in which some words have single-stroke notations, uses a similar approach, and while shorthand is great for writers with fast-talking bosses, it is notoriously hard to read. Kids also don't seem to think this is a good idea—there's not a single lone line in all of my daughter's drawing in Figure 1. One reason it's not a good idea is that there just are not enough distinguishable types of strokes to cover all the words we speak. Coming up with even a hundred easily distinguishable stroke types would be tricky, and we'd need tens of thousands for reasonably efficient writing.

There is also a more fundamental difficulty with using single strokes, and it has to do with the way our brains work. The part of our brains that performs visual computations is arranged in a hierarchy. The lower areas of the hierarchy—the first place where visual processing takes place on the brain's way to object recognition—deal

"The rain in Spain stays mainly in the plain."

a. The sentence with strokes for words

b. The sentence with objects for words

c. The sentence with scenes for words

FIGURE 3. *Three different strategies for writing, here used to write the spoken sentence about the rain in Spain.* (**a**) *One could use single strokes for every word, leading to writing akin to that shown: short, but with words that don't look object-like.* (**b**) *Alternatively, one could use object-like symbols for the meaningful words, as shown (and still use single strokes for "function" words like "the," "in," and "mainly"). This is the strategy many cultures have come to use because it best harnesses our visual system's object-recognition abilities for reading.* (**c**) *A third strategy would be to use drawings more complex than objects to write spoken words, which is not useful because then the objects our visual systems identified within those words (the shield-shape in "rain," for example) would not have any specific meaning in and of themselves.*

with simpler parts like contours, slightly higher areas then deal with simple *combinations* of contours, and, finally, the highest areas of the hierarchy recognize and perceive full objects. The problem with using single strokes to represent spoken words, as in Figure 3a, is that the visual system finishes processing the strokes far too early in the hierarchy. The visual system is not accustomed to creating word-like (e.g., object-like) interpretations from single strokes. We don't typically perceive single strokes at all, at least not in the same way we see objects. For example, when you look at Figure 4 you perceive a cube in front of a pyramid. *That's* what you consciously notice and judge. You don't "see" the dozen-plus contours the same way. Nor do you "see" the many corners and junctions where those contours intersect. You don't say, "Hey, look at all those contours and corners in the scene." Our brains evolved to perceive objects, not object-parts, because objects are what stay connected over time and are crucial to parsing and making sense of the world. Our brains naturally look for objects and want to interpret outside stimuli as objects. So, using a single stroke for a word (or using a junction for a word) does not make our brains happy. Instead, when seeing the stroke-word sentence in Figure 3a, the brain will desperately try to see objects in the jumble of strokes. If it can find one, it will interpret that particular jumble of strokes in an object-like fashion. But if it did this, it would be interpreting a phrase or whole sentence as an object, and that's not helpful for understanding a sentence. Using single strokes as words is a bad idea, then, because the brain is not designed to treat single contours as meaningful. The brain is also not designed to treat object junctions as meaningful. That's why spoken words tend to be written with symbols with complexity similar to that of visual objects.

What if, instead, we visually symbolized spoken words with whole scenes, i.e., via multiple objects rather than just a single one? Figure 3c shows what "The rain in Spain..." might look like using this "scene-ogram" writing. Quite an eyeful. The drawings found in some furniture assembly manuals follow this strategy. The problems now are the opposite of the ones we had before. First, each scene-ogram image reads more like a sentence than a word—like "Take the nail that looks like this, and pound it into the wooden frame that looks like that." Secondly, having object-like symbols as part of these complex symbols is itself a problem because it makes the brain interpret these object-like symbols as objects with their own meanings. However, these object-like symbols are now just the building blocks, parts of a written word that have no meaning in and of themselves at all.

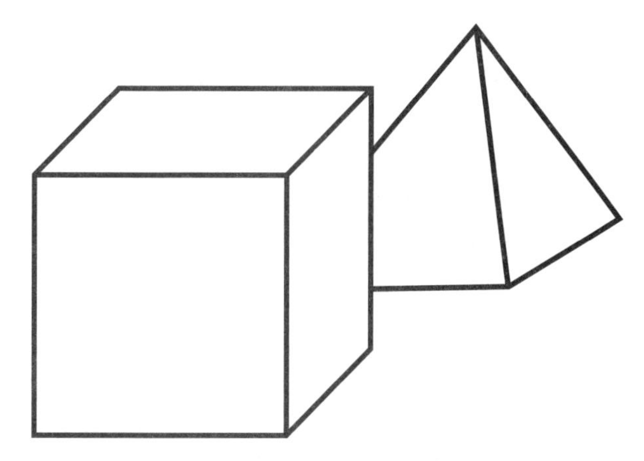

FIGURE 4. *When you see this figure, you see a box partially occluding a pyramid. That is, you see objects. You don't see fourteen contours. You don't see twelve junctions (places where contours intersect). Writing has culturally evolved to take advantage of what our visual system has, over hundreds of millions of years, evolved to do: see objects. Writing has evolved so that words tend to look object-like.*

The visual system possesses innate mechanisms for interpreting object-like visual stimuli as objects. Because spoken words are the smallest meaningful entities in spoken language, and often have meanings that are at the object level (they are either objects, or properties of objects, or actions of objects), it is only natural that to represent them, we use stimuli that our visual system has been designed to not only interpret, but interpret as objects. By using objects to stand for spoken words—and not smaller-than-object visual structures like lines or junctions, and not larger-than-object visual structures like scenes—we are best able to harness the visual system's innate abilities, even as we are asking it to perform a task it never evolved to do (see Figure 5).

Object-like symbols might, then, be a good choice to represent words, but are these object-like symbols a result of cultural evolution, having been selected for, or might these symbols be object-like only because the first symbols, historically, were object-like? After all, the first symbols tended to be object-like pictograms, even more object-like than the symbols in Figures 2b and 2c. Perhaps our symbols are still object-like merely because of inheritance, and not because they have been designed to be easy on the eye. The problem with this argument is that writing tends to change quickly over time, especially as cultures split. If there were no pressure from cultural selection to *keep* symbols looking object-like, then the symbol shapes would have changed randomly over the centuries, and the object-likeness would have tended to disappear. But that's not what we see when we look at the history of written language. Culture has seen to it that our symbols retained their object-likeness, because that's what makes us such good readers. It *is* interesting, though, that even the first symbols were on the right track, before cultural evolution had time to do any shaping. Although, given that even small children manage to cotton on to this, perhaps it's not too surprising that the first scribes appreciated the benefits of object-like drawings for words as well.

The Trouble with Speech Writers

The brain prefers to see objects as the symbols for words, and kids and much of the rest of the world have complied. Such writing is "logographic"—meaning it employs symbols for words rather than for sounds—and doesn't give the reader information on how to speak it. This is actually a great benefit, because it means that even people who speak different languages can utilize the same writing system and be able to communicate using it. That is, because the symbols in logographic writing systems indicate concepts rather than sounds, they can serve as universal writing systems, bringing together a variety of spoken languages into harmony and friendship. Japanese speakers, for example, have no idea what a Chinese speaker means when he or she is speaking, but can understand a fair bit of written Chinese because Japanese speakers also use Chinese characters.

Brotherhood and peace are nice, but there er jus some thangs ya cayant do when writin' with objects... including putting a person's accent down on the page the way I just did there. Someone who can read Chinese may still be totally unprepared to speak to anyone in China, because knowing the logographic symbols isn't the same thing as knowing what the words sound like. The kind of writing you're reading right now is entirely different. Rather

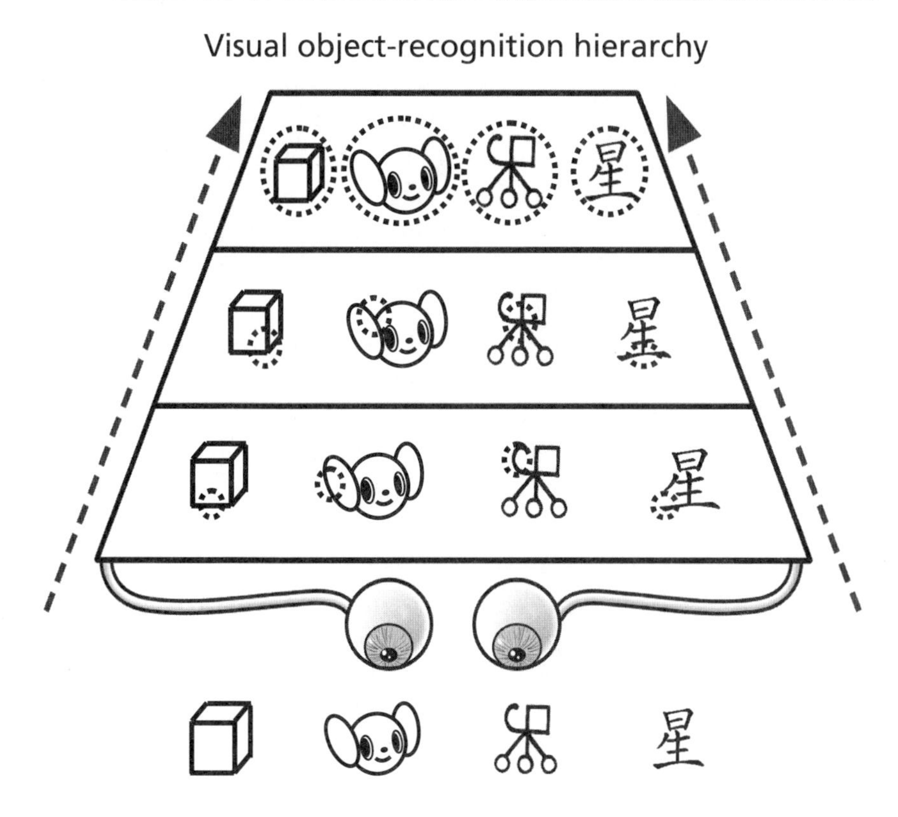

FIGURE 5. *The part of the human visual cortex responsible for recognizing objects is organized into a hierarchy, where the lower levels are responsible for recognizing simpler visual features of objects such as edges or strokes, middle levels are responsible for simple contour combinations like junctions, and the highest levels are responsible for whole objects. For simplicity, I've shown this as just three levels; in reality, there are about a dozen. Ovals on each drawing show the level of detail that each part of the brain deals with. Whether we're dealing with real-world objects, semi-symbolic cartoons, visual signs outside of language proper, or symbols in logographic writing like Chinese, symbols in writing tend to be roughly object-like—just what our visual system evolved for.*

than symbols for objects, the basic symbols are letters that tell you how to speak words that refer to objects. You're reading "speech-writing." Speech-writing allows us to put Tom Sawyer's accent on paper, and it allows someone who doesn't speak our language to obtain a significant amount of knowledge about how to do so just by reading at home. Such a person would have an atrocious accent, but they'd be off to a great start. A second important advantage to speech-writing is that one can get away with many fewer symbols. Rather than one object-like symbol for each of the *tens of thousands* of commonly spoken words, one only needs a symbol for each of the *dozens* of speech sounds, or phonemes those words contain. That's a thousand-fold reduction in the number of written symbols we have to learn.

I have no idea whether the merits of speech-writing (focus on pronunciation, simplicity) outweigh the benefits of logographic writing (direct one-to-one object-to-symbol correlation, easier communication across spoken languages), but there *have* been hundreds of speech-writing systems throughout history, and many are still in use today, by about half the world's population. And when culture decided to go the speech-writing route rather than just stick with logography, it created a big dilemma. As we've discussed, the best way to harness the natural object-recognition powers of the visual system is to have spoken words look object-like on paper. But in speech-writing the symbols stand for speech sounds, and written words almost always consist of *multiple* speech sound symbols. How can our written words look like objects if written words can no longer be associated with specific object-like symbols? If symbols stand for speech sounds rather than objects, then the look of a written word will depend upon the letters in it. That is, the word's look will be a result of how the word *sounds* when spoken, rather than being designed specifically to look like an object. Had the word been spoken differently, the written word would look different. And if the look of a word depends on how speakers *say* the word, rather than each word being given its own special visual symbol, it would seem that all hope is lost in trying to make written words look object-like in speech-writing.

There is a way out of this dilemma, however. And although no one individual may have originally come up with it, culture has nevertheless evolved to utilize it. The way out is this: *If written words must be built out of multiple symbols, then to make words look object-like, symbols should look like object* **parts**. And as we'll see, that's what culture did. Culture dealt with the speech-writer dilemma by designing letters that look like the object *parts* found in nature. That way written words will typically be object-like, and our visual system will still be effectively harnessed for the unnatural act of reading. Before getting to the *shapes* of letters, in the next section I'll begin by asking whether letters are the right size—i.e. whether they have the right number of strokes—to be object parts.

Junction Conjunction

Letters need to look like object parts if there's any hope that the words they create will look like objects. At a minimum, then, letters should be smaller than objects, i.e., involve fewer strokes than are typically found in an object. But how small should letters be? It depends on how many distinct speech sounds, or phonemes, are in a typical word. For English, as in many languages, the average number of phonemes per spoken word is about four or five. There are many words that have more phonemes than this, but we rarely use them and so they don't affect the average much. In speech writing, phonemes tend to correspond closely to letters (though they don't always, as with the phoneme "th"). If we want written words to be object-like, then, we don't want to use single strokes for letters because then the typical word would have only around four or five strokes, arranged side by side on the page, and fewer strokes in writing systems where not every phoneme corresponds to a single letter. The visual appearance of the resulting words would fail to be object-like in two respects. First, the number of contours would usually be lower than that found in a typical object. You can't draw much of an object with four or five strokes; a simple cube requires nine just on its own. Second, the resulting word would be lacking the internal structures expected of an object. Real objects in real scenes don't have contours that simply lie next to one another. Instead, contours in natural scenes interact, or touch, one another in characteristic ways. For example, take another look at the cube and pyramid in Figure 4, and note all the places contours meet. Single-stroke letters wouldn't give us object-like words, then, because the resulting words would be short on contours and internal structure.

A solution would seem to suggest itself: Put a little "structure" into the letters—more than a single stroke, but fewer strokes than an object requires. There are fundamental structures in between contours and objects, structures long known by engineers and vision scientists for their importance in parsing images into objects. Some of these structures are "junctions," the places where contours intersect at a point, such as in the Ls, Ts, and Ys in Figure 4. Others are simple concatenations of junctions. By using such configurations of strokes as the written symbols for letters, we would alleviate both of the problems with single-contour letters. First, if each letter tends to have a few contours, then a typical written word of four or five letters would have about a dozen strokes, enough to handle the simple cartoon animals in Figure 2a. Second, the written words would then include the internal structures we expect to see in images of objects. Making letters configurations of contours rather than single contours leads to words that are much more object-like—but not *entirely* object-like, because the manner in which the junctions themselves are arranged on the page will be different than in nature. The junctions will be arranged in an unnatural sequence, and two junctions may end up next to each other that would never be so close in nature. Nevertheless, this solution is the best we can hope for in speech-writing systems.

Have speech-writing systems employed this way out of their dilemma? That is, do letters tend to have several strokes? Yes, in fact—a point I first made, with my co-author Shinsuke Shimojo, in a 2005 publication in the *Proceedings of the Royal Society of London*. I measured the average number of strokes per letter in the ninety-three speech-writing systems listed in Figure 6 and found that the average tends to be three no matter how many characters a writing system has, fifteen or a hundred fifty. Rather than having more strokes per letter, writing systems with more characters use more *kinds* of strokes to build their letters, keeping the number of strokes per letter constant at around three.

As expected, then, letters *are* a bit like object *parts*, at least in size (i.e., number of strokes). Written words therefore end up looking more object-like, allowing us to read more ably. There is a way for words to look even *more* object-like, however: if letters weren't merely roughly the right size for intermediate object parts, but were actually composed of the same *shapes* as the object parts found in

Name, and sample characters		Average number of strokes per letter	Total number of letters
Ahom	[illegible]	2.40	40
Albanian (Elbasan)	[illegible]	2.60	53
Ancient Berber (Vertical)	[illegible]	2.68	25
Arabic	ط ء ك	2.63	35
Arabic	١ ٢ ٣	2.00	10
Aramaic	[illegible]	2.68	22
Armenian (Eastern)	[illegible]	2.21	39
Asomtavruli	[illegible]	2.42	38
Avestan	[illegible]	2.42	53
Bassa	[illegible]	2.37	30
Batak (Kara Batak)	[illegible]	2.15	33
Bengali	ক খ গ	3.91	45
Bengali	১ ২ ৩	1.80	10
Brahmi	[illegible]	2.14	43
Buhid (Mangyan)	[illegible]	3.78	18
Burmese	က ခ ဂ	2.51	41
Burmese	၁ ၂ ၃	1.40	10
Carrier (Dene)	[illegible]	3.20	180
Celtiberian	[illegible]	3.29	28
Cherokee	[illegible]	2.35	85
Chinese	一 二 三	3.00	10
Cypriot	[illegible]	3.84	55
Cyrillic (Abkhaz)	ь э ю	3.69	62
Dehong	[illegible]	3.39	33
Dehong	[illegible]	1.90	10
Deseret	[illegible]	1.68	38
Devanagari	क ख ग	3.27	45
Devanagari	१ २ ३	1.40	10
Dives Akuru	[illegible]	3.77	23
Enochian	[illegible]	2.52	21
Ethiopic (Ge'ez)	ሀ ለ ሐ	2.63	40
Etruscan (archaic)	[illegible]	2.91	23
Faliscan	[illegible]	2.52	21
Fraser	[illegible]	2.37	41
Glagolitic	[illegible]	4.51	41
Gothic (Wulfila)	[illegible]	2.32	25
Greek	α β γ	1.71	24
Gujarati	ક ખ ગ	2.21	42
Gujarati	૧ ૨ ૩	1.50	10
Gurmukhi	ਕ ਖ ਟ	3.20	46
Gurmukhi	੧ ੨ ੩	1.90	10
Hanuno'o (Mangyan)	[illegible]	3.13	16
Hebrew	ט ה ו	2.55	33
Hindu-Arabic	1 2 3	1.60	10
Hungarian Runes	[illegible]	3.08	40
Hungarian Runes	I II III	2.50	6
Iberian (northern)	[illegible]	3.46	26
Iberian (southern)	[illegible]	3.14	22
International phonetic	ʃ ɛ ʕ	2.42	170
Kannada	[illegible]	2.79	47
Kannada	೧ ೨ ೩	1.00	10
Kharoshthi	[illegible]	1.72	39
Kharoshthi	[illegible]	1.88	8
Khmer	[illegible]	7.49	68
Khmer	១ ២ ៣	3.70	10
Korean (Hangeul)	ㅏ ㅛ ㅈ	2.83	24
Kpelle	[illegible]	3.07	88
Latin, ancient	A B C	2.67	21
Latin, modern	a b c	2.08	26
Latin, modern all-caps	A B C	2.50	26
Lepcha (Rong)	[illegible]	2.68	77
Lepcha (Rong)	[illegible]	2.60	10
Limbu	[illegible]	2.51	37
Linear B	[illegible]	5.03	73
Marsiliana	[illegible]	2.88	26
Meroitic (non-hieroglyphic)	[illegible]	3.46	23
Messapic	[illegible]	2.87	23
Middle Adriatic (South Picene)	[illegible]	2.70	23
Middle Persian (Pahlavi)	[illegible]	2.00	22
Mkhedruli	ა ბ გ	2.21	38
Mongolian	[illegible]	3.29	35
Mongolian	[illegible]	1.60	10
Nabataean	[illegible]	1.77	22
Ndjuka'	[illegible]	2.35	52
New Tai Lue	[illegible]	2.23	10
N'Ko	[illegible]	2.14	22
N'Ko	[illegible]	2.60	10
North Picene	[illegible]	2.67	18
Nuskha-khucuri	[illegible]	3.58	38
Old Church Slavonic	[illegible]	3.33	42
Old Permic (Abur)	[illegible]	3.11	38
Oriya	[illegible]	2.89	44
Oriya	[illegible]	1.50	10
Oscan	[illegible]	2.71	21
Pahawh Hmong	[illegible]	3.30	86
Pahawh Hmong	[illegible]	2.50	10
Parthian	[illegible]	2.59	22
Pashto	ځ ښ ډ	2.73	40
'Phags-pa	[illegible]	4.53	40
Phoenician	[illegible]	2.86	22
Pollard Miao	[illegible]	2.11	47
Psalter	[illegible]	2.00	21
Redjang (Kaganga)	[illegible]	3.00	36
Runic (Danish Futhark)	ᚠ ᚢ ᚦ	2.63	16
Runic (Elder Futhark)	ᚠ ᚢ ᚦ	3.13	24
Sabaean/Minean	[illegible]	3.55	29
Samaritan	[illegible]	4.32	22
Santali (Ol Cemet')	[illegible]	3.33	30
Santali (Ol Cemet')	[illegible]	1.40	10
Sil'oti Nagri	[illegible]	3.33	34
Somali (Osmanya)	[illegible]	1.90	30
Somali (Osmanya)	[illegible]	1.50	10
Sorang Sompeng	[illegible]	2.29	24
Sorang Sompeng	[illegible]	1.40	10
South Arabian	[illegible]	3.29	28
Soyombo	[illegible]	3.63	35
Syriac	[illegible]	2.27	22
Tagalog	[illegible]	1.93	16
Tagbanwa	[illegible]	2.23	15
Tamil	க ங ஐ	2.74	34
Thaana	[illegible]	2.09	35
Theban	[illegible]	3.83	24
Tifinagh	ⵣ ⵉ ⵅ	2.88	25
Umbrian	[illegible]	2.48	21
Varang Kshiti	[illegible]	2.86	21

FIGURE 6. *A table of the ninety-three speech-writing systems for which I measured the average number of strokes per letter. The table shows sample letters, the average number of strokes per letter, and the total number of letters the writing system contains. A nice Web site for seeing many more example letters is Simon Ager's www.omniglot.com.*

nature. As we'll see, only *certain* contour-configurations tend to be found in nature. Once we know which configurations these are, we can then ask whether letters have *these* shapes. That is, we'll then be able to decide whether letters, and therefore the words they form, look like objects in nature.

Natural Subthings

The shapes L and V look different, but in another sense they look the same. Each has two strokes, joined at their ends, and thus the two have the same basic configuration. Rotate, stretch, and twist these letters as much as you wish, but you'll never get a T, nor will you get an X, Y, K, Ψ, F, or π (see Figure 7a). All these stroke configurations—just some of the configuration types possible—are distinct from one another. *This* is what I mean when I refer to a letter's shape (really a kind of topological shape), and it is just the notion of shape we need to use as we look for similarities between letters and configurations in nature. In particular, we'll be interested in the configurations with up to three strokes and two intersections, namely those shown in Figure 7b, a kind of "periodic table" of possible shapes I call "subthings"—parts that are used to make up objects. We want to know which of these shapes are found in nature. But before we look at the shapes found in images of nature, let me introduce you to some of these shapes, and give you a qualitative impression of how common they are.

Let's begin with one of the simplest kinds of subthing: L-junctions. The natural "habitat" of L-junctions is corners. Namely, the corners of objects, like in the left-hand image in Figure 8a, where you see only one of the object's faces at that corner. Because corners are common in the world, Ls are common, and so we'll expect to find lots of L-junctions among letters when we look at them later on. Next up is the T-junction, which occurs when one contour's endpoint buts up against the side of another contour. T-junctions are typically caused by partial occlusion, that is, when one object is behind another, as shown in the left-hand image in Figure 8b. Like L-junctions, these kinds of junctions are very common because we are typically surrounded by lots of opaque objects, some of which are in front of

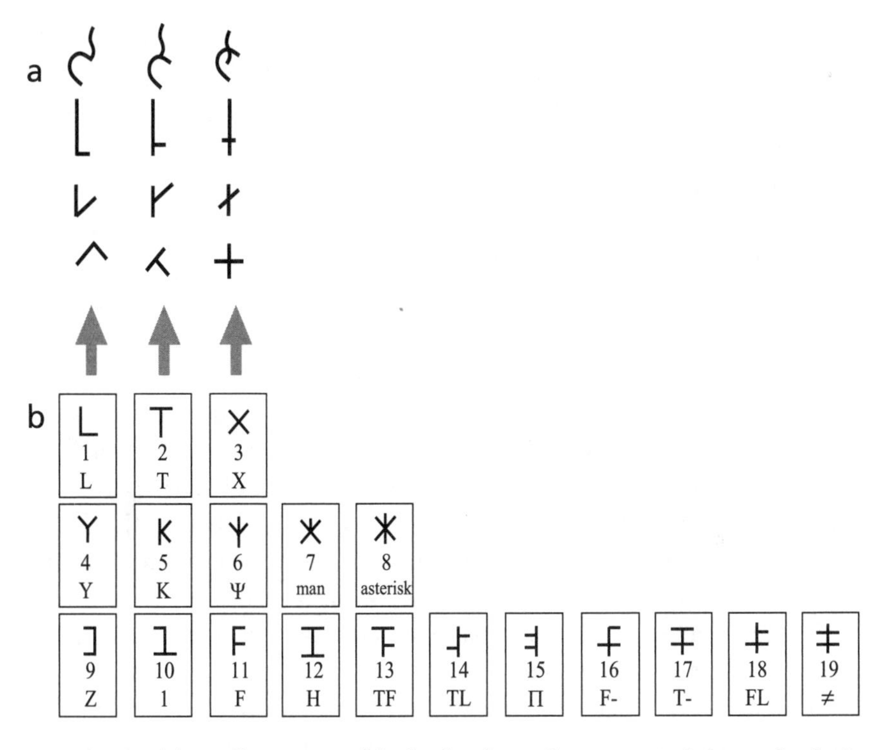

FIGURE 7. (**a**) *An illustration of the kinds of specific geometrical shapes for the L, T, and X configurations. Each configuration keeps its shape even if the entire configuration is rotated, the angle of intersection is changed, stroke lengths are modified, or the shapes of the strokes changed (so long as the stroke remains smooth).* (**b**) *The "periodic table" of all nineteen configurations using two or three contours and one or two junctions (intersection points).*

the others. We'll therefore expect to find that speech-writing systems have a lot of these junctions.

The third and last configuration built from two contours is X, which is created where contours cross. Compared to L- and T-junctions, X-junctions are not as easy to find in nature. Look around; I'll bet you can't find many around you now. A common mistake is to think that two sticks or pencils that cross each other make an X-junction. If they're far enough away that the sticks look like a single contour, then the result would indeed be an X-junction. But if the contours on either side of the sticks are visually distinguishable, then one stick crossing behind another is a case of partial occlusion, actually making four T-junctions and no X-junctions. Where we *do* find X-junctions are in cases of partial transparency, e.g., when a piece of tinted glass is in front of some other object, as illustrated in the left-hand image in Figure 8c. Needless to say, tinted glass is not something our ancient ancestors saw much. The only other situation where we find X-junctions is when there are stacks of objects, like the bricks in the right-hand illustration in Figure 8c, where the contours of the X are actually the *cracks* between objects—i.e., when the edges of two objects lie atop one another so closely that the space between them appears to be a single contour. This is also rare compared to L- and T-junctions. And L- and T-junctions can *also* be due to cracks, as shown in the right-hand images in Figures 8a and 8b. X-junctions are a rarity, then, in nature, so we'll expect to find few of them in writing. For written English, at least, we know already that this is the case, because "X" is used so rarely that we can readily remember the two most common words it begins: X-ray and xylophone.

Figure 9 covers the next batch of configuration types, each of which is created from three contours that intersect at the same spot. It includes Y, K, Ψ, a stick-figure-shaped junction, and an asterisk-shaped junction. Y-junctions are the most common among these, occurring at the corners of objects where three faces are visible. As can be seen in Figure 9, the other configurations require successively greater coincidences in the way objects are arranged. This means that they become progressively rarer.

We now have a rough impression of where the configurations in the first two rows of the periodic table of shapes in Figure 7b can be found in nature, and how common they are. The third row of Figure 7b shows configurations with two points of intersection, where each is a combination of two of the simpler configurations in the first row of Figure 7b, i.e., combinations of Ls, Ts, and Xs. It is instructive to consider four of these in more detail, namely #12 through #15. Each of these is built out of two T-junctions. Recall from Figure 8b that the typical cause of a T-junction is partial occlusion, so we'd expect to find that each of these four cases could be found in cases where objects are behind other objects. Figure 10 shows some simple scenes with partial occlusions that do indeed lead to these configurations... except in the case of #14, where one contour pokes

a. L-junctions: common

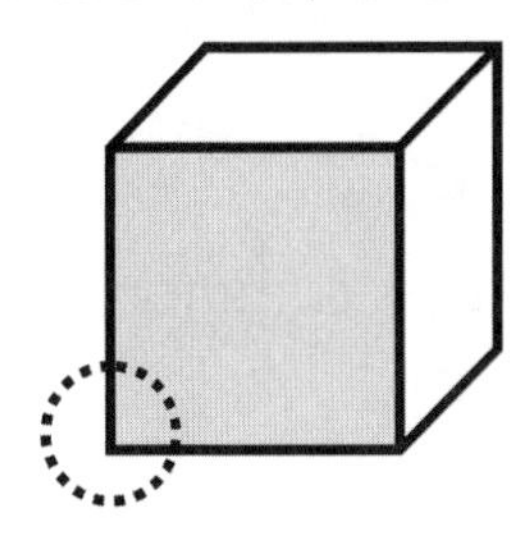

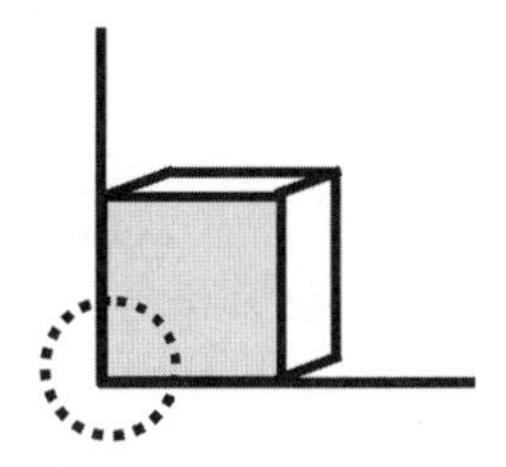

b. T-junctions: common

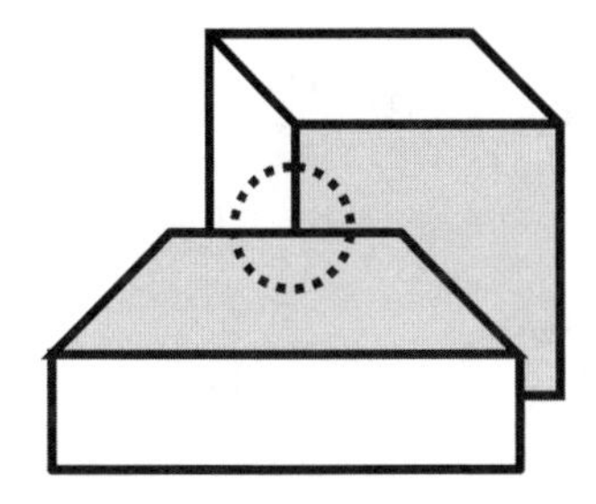

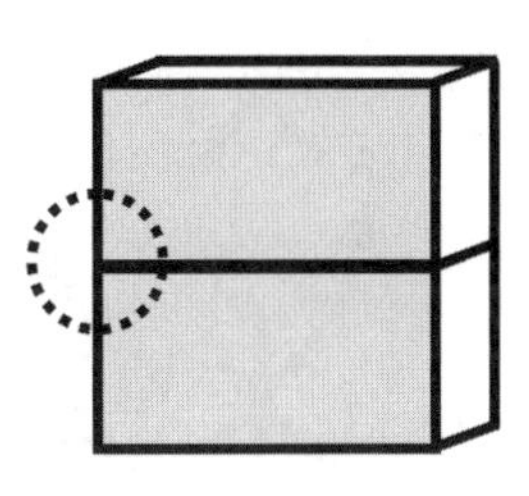

c. X-junctions: rare

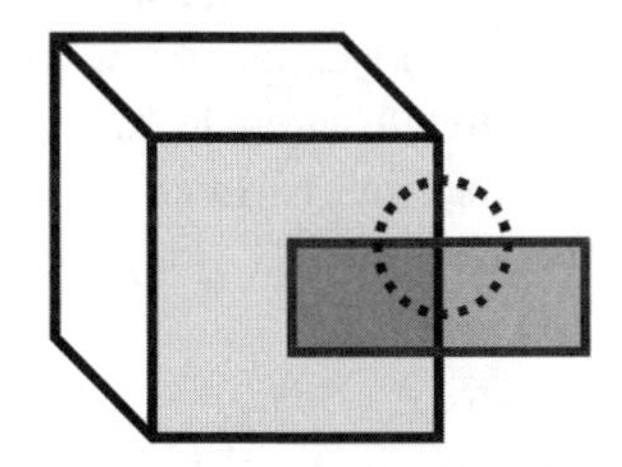

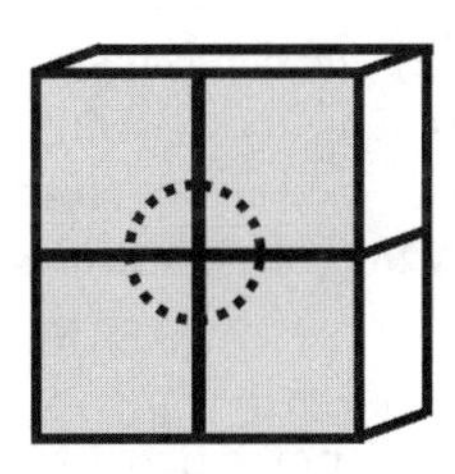

FIGURE 8. *Different types of junctions occurring in nature.* (**a**) *L-junctions are typically caused by corners where only one face is visible. Much more rarely, they can also be caused by cracks between objects, such as when a block is placed in a corner.* (**b**) *T-junctions are typically caused by partial occlusion, where a contour goes behind another. But Ts can also be created from cracks, like when one brick is laid on another brick.* (**c**) *X-junctions don't occur from simple arrangements of opaque objects. They can occur, however, when an object is seen through tinted glass, as shown. And they can also occur from cracks, as in stacks of blocks. Each of these arrangements of objects is rare, and so Xs are rarer than Ls and Ts in nature, something we'll see is the case in Figure 11.*

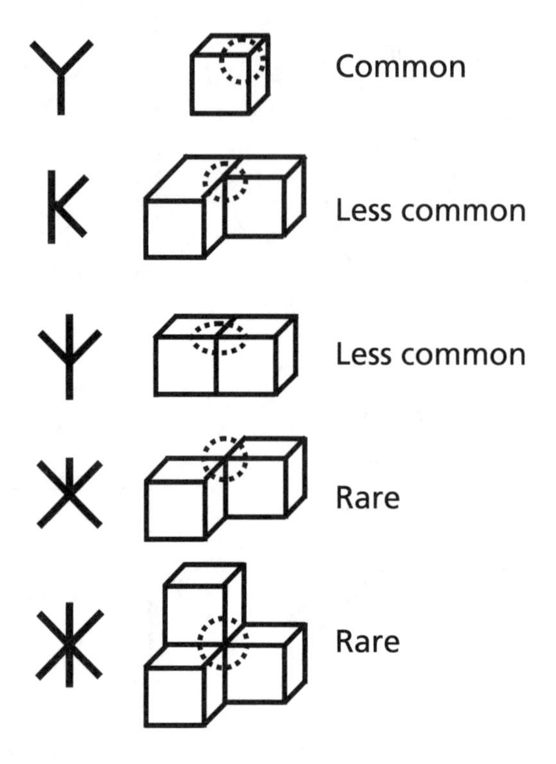

FIGURE 9. *The five kinds of junctions created by three contours: Y, K, Ψ, "man," and "asterisk." An illustration of the kind of object arrangement required to get each kind of junction is also shown, along with an indication of how common that junction is. Roughly speaking, as you move from top to bottom in the figure, greater coincidences in the way objects are arranged are required to create the junction types. (Figure 11 demonstrates that this is indeed the case.)*

out from either side of the center contour. Unlike his three cousins, this configuration *cannot* occur due to partial occlusion. You can picture this yourself by imagining that one of the poking-out contours is the edge of an object that is partially occluded by the vertical line. However, if this is the case, then the object in front, the one doing the occluding, is on the other side of the vertical line. *That* makes it impossible to create the other segment that pokes out from partial occlusion, because that other segment would have to be a contour *on* the object for the configuration to be formed. There are scenarios where configuration #14 *can* occur, such as in the stack of bricks

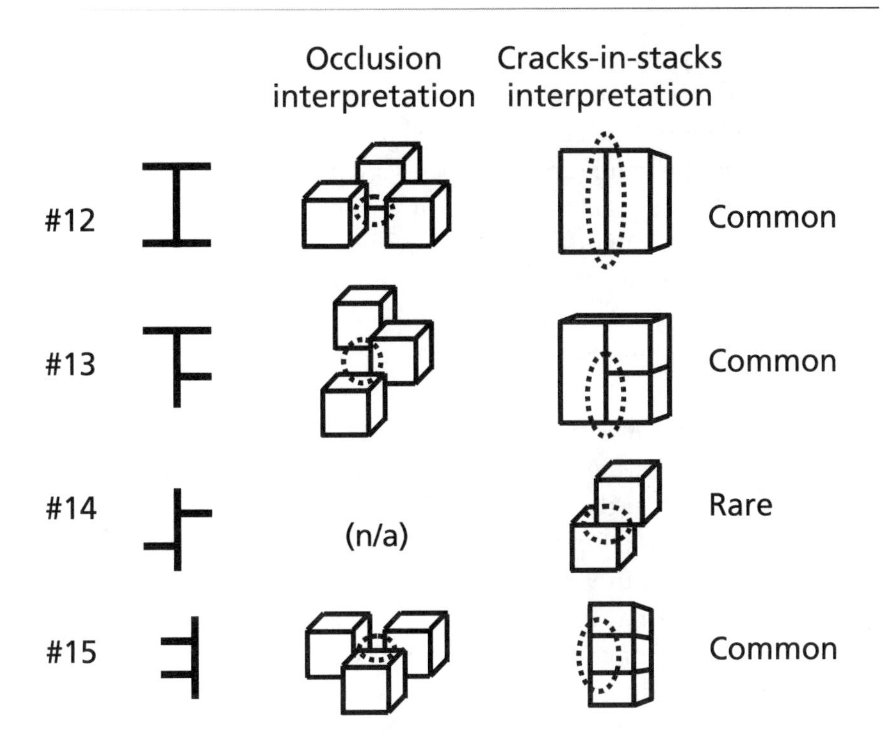

FIGURE 10. *The four contour configurations built using two T-junctions. The main cause of T-junctions is partial occlusion, but there is no possible way to get the third configuration through occlusion. The third configuration type can, however, occur in the cracks in stacks of blocks, just as the other three configuration types can. This means that the third configuration type above must be rare in images relative to the others, something you can see is the case in Figure 11.*

shown in Figure 10, but there are *also* ways to stack bricks to acquire the other three two-T cases. This means that #14 is rare in the world compared to the other three configurations. Therefore, we should expect it to be rare in speech-writing. As we'll see later, this is indeed the case.

So far, I've been making intuitive arguments for how common these configurations are. But this only helps us get a *qualitative* sense of their relative commonness, and only for a fraction of these shapes. To find out quantitatively how common the nineteen shapes are I had two undergraduate Caltech students, Qiong "Gus" Zhang and Hao Ye, spend a summer—and more—measuring the frequency of configuration types in images. Some of the images were from tribal scenes, some were from *National Geographic*, and some were from computer-generated images of commercial parks. Figure 11 shows how common the configuration types were in each of the three classes of image, and as you can see, the three plots look nearly the same. How well do these quantitative measurements match the conclusions of the intuitive arguments earlier in this section? We saw then that L- and T-junctions were common but X-junctions comparatively rare, and we see this in Figure 11 as well. We also saw that among Y, K, Ψ, "man," and "asterisk," Y-junctions should be most common, and the others should become progressively rarer. Again, Figure 11 shows that this is the case. Finally, recall that our thought experiment led us to the conclusion that #14 should be rare compared to #12, #13, and #15. Once more, Figure 11 shows that this is the case.

The information in Figure 11, first published in a 2006 paper by myself, Qiong Zhang, Hao Ye, and Shinsuke Shimojo in *The American Naturalist*, summarizes what nature looks like, at the level of the shapes of subthings found in the world; it shows the "signature" of nature. Now that we have this information, we can compare the frequency of these subthings to the frequency of subthings found in speech-writing. In the next section, we'll do just that: we'll see if letters have the same shapes, in the same frequency, as nature. But before we do this we might want to check a claim I made earlier: that culture has created visual signs and the symbols in logographic writing to look object-like. Although this intuitively appears to be the case—see Figure 2 again—now that we know the subthing shapes of nature we can finally see whether visual signs that seem object-like really are object-like, by determining whether the subthing shapes we see in visual signs are the ones we also see in nature. Figure 12b shows the frequency of subthing shapes in Chinese characters and a large set of non-linguistic signs, and as you can see, they possess the same natural signature we saw in Figure 11 (reproduced in Figure 12 as Figure 12a for easy comparison).

That visual signs and symbols for both objects and words have culturally evolved to look object-like is a fundamental observation, and crucial to understanding why logographic writing is so effective. However, it is not utterly surprising because these symbols do *seem*

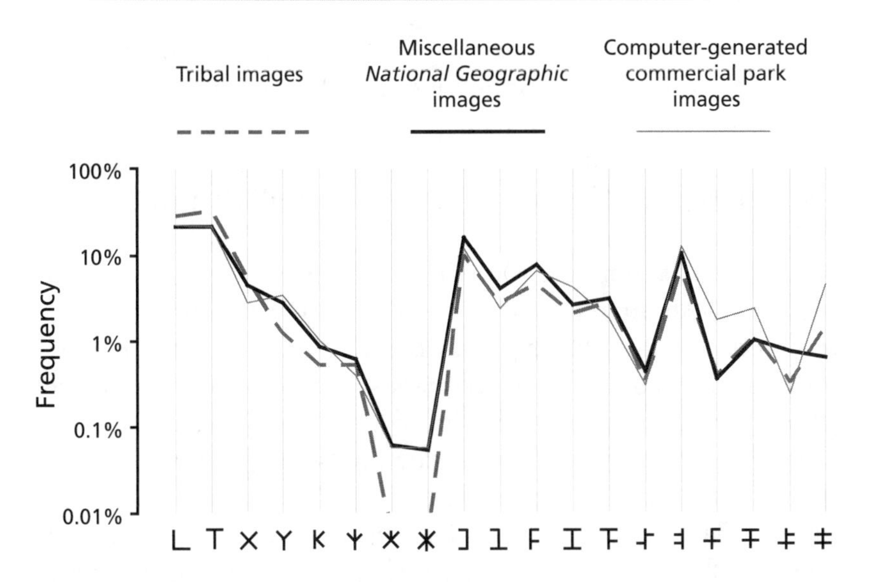

FIGURE 11. *How common the nineteen configuration types are in three kinds of natural images: tribal, miscellaneous* National Geographic, *and computer-generated images of commercial parks. Notice how similar the distributions are despite the different kinds of image. Thanks go to Qiong Zhang and Hao Ye for their hard work helping to acquire some of these data while they were Caltech undergraduates.*

object-like. The next question we must ask is whether these frequencies also occur in speech-writing, where the expectation is for letters to be shaped like natural subthings, so that words will—like visual signs and logographic writing—look object-like. That's where we'll turn now.

Finding Nature in Letters

There is something alluring about finding letters in nature ("Hey, that tree trunk over there ducks behind that other trunk in just the right way to create a K!"). Kjell Sandved has accumulated gorgeous photographs of the Latin letters in nature, and even an entire set of letters on the wings of butterflies. Krystina Castella and Brian Doyle published a variety of images of letters in nature, and on my way back from having lunch with them one day I ran into a Caltech colleague who informed me that *he* had a long interest in trying to find letters in the contours of nature. An artist in the Los Angeles area once contacted me and informed me that she'd been onto the connection between letters and nature for thirty or more years, having discovered it by spending an entire summer staring in meditation at the reeds in a pond. She said she'd also observed letter shapes even in the shapes of neurons!

Finding letters in nature is fun, but the mere fact that we can play this game *at all* is an important observation. Consider some less popular varieties of this "finding X in nature" game: "finding Braille," "finding plaids," and "finding ear-shapes." These games would never catch on, because it's just too hard to find Braille, plaids, and ear-shapes in nature. Letters, however, are pretty easy, and *that* suggests that the reason may be because letters are shaped like nature. That is, the very fact that we play this letter-game hints at a fundamental technological insight: that letters are shaped like natural subthings. A simple swapping of words—from "finding letters in nature" to "finding nature in letters"—takes us from a game to a discovery.

But do we actually find nature in letters? We know that letters are built from roughly contour-like strokes, and that they look more like nature than they do, say, photographs of knots. And we also know from earlier that the number of strokes per letter is about three, making letters just the right size to be subthings. What we haven't yet determined is if letters have the same *shapes* as natural subthings. Figure 13b shows the frequency with which the various configurations we've been discussing occur in about 100 speech-writing systems across human history. You can immediately see its close similarity to the frequencies plotted in Figure 13a (the same graph as in Figure 11 and Figure 12a). Speech-writing reflects the same signature of nature. That is, nature really *is* in the letters of speech-writing! And *this* lets written words in speech-writing look more object-like, which is what enables us to so effectively harness our visual system for reading, as Figure 14 shows. What this tells us about human visual signs generally, then—whether non-linguistic symbols, logographic writing like Chinese characters, or speech-writing—is that they possess substructures shaped like natural subthings, and that they are therefore shaped so that words appear object-like. Figure 15 shows the composite rank order for configuration frequency in nature ver-

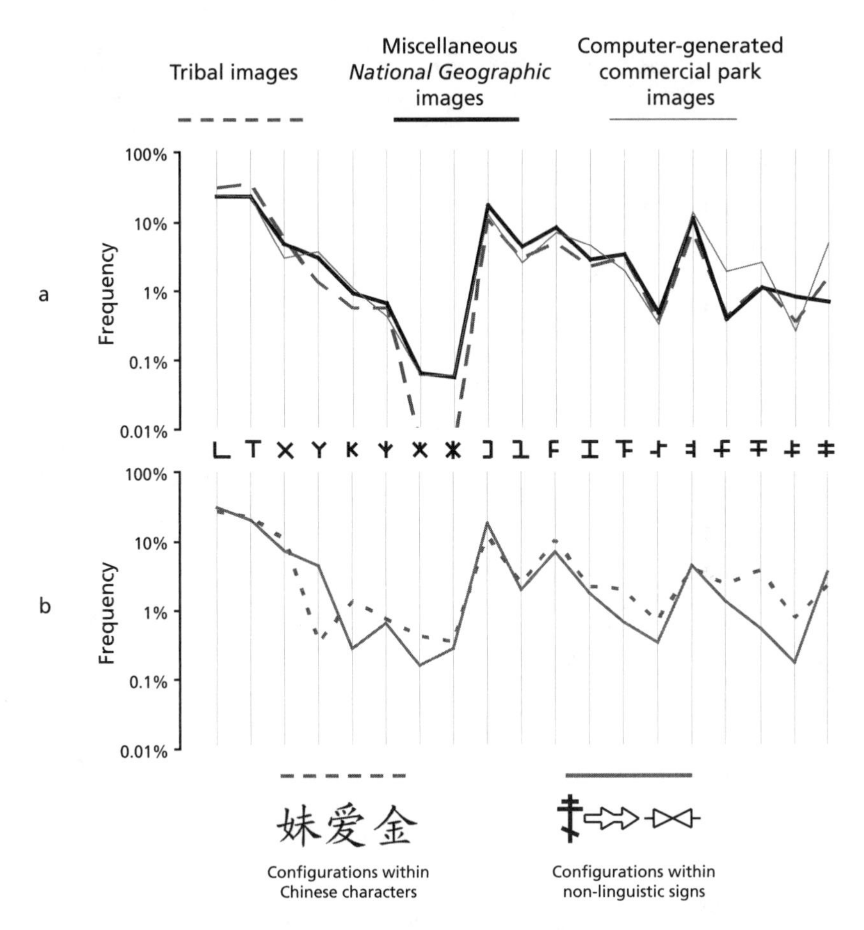

FIGURE 12. (**a**) *The frequency distribution of subthings in natural images, repeated from Figure 11.* (**b**) *The distribution of configurations found within the shapes of Chinese characters and non-linguistic visual signs. You can see that the configurations found in these logographic symbols is very similar to those found in natural images. This is somewhat surprising because most of these symbols don't look like specific objects, but it is not overly surprising because these symbols do tend to look roughly object-like. Thanks go to Qiong Zhang for his hard work helping to acquire some of these data while he was a Caltech undergraduate.*

sus human visual signs of all kinds. In general, the more common configurations in nature are also the more common ones in human visual signs.

The natural signature we've been looking at will hold true any-

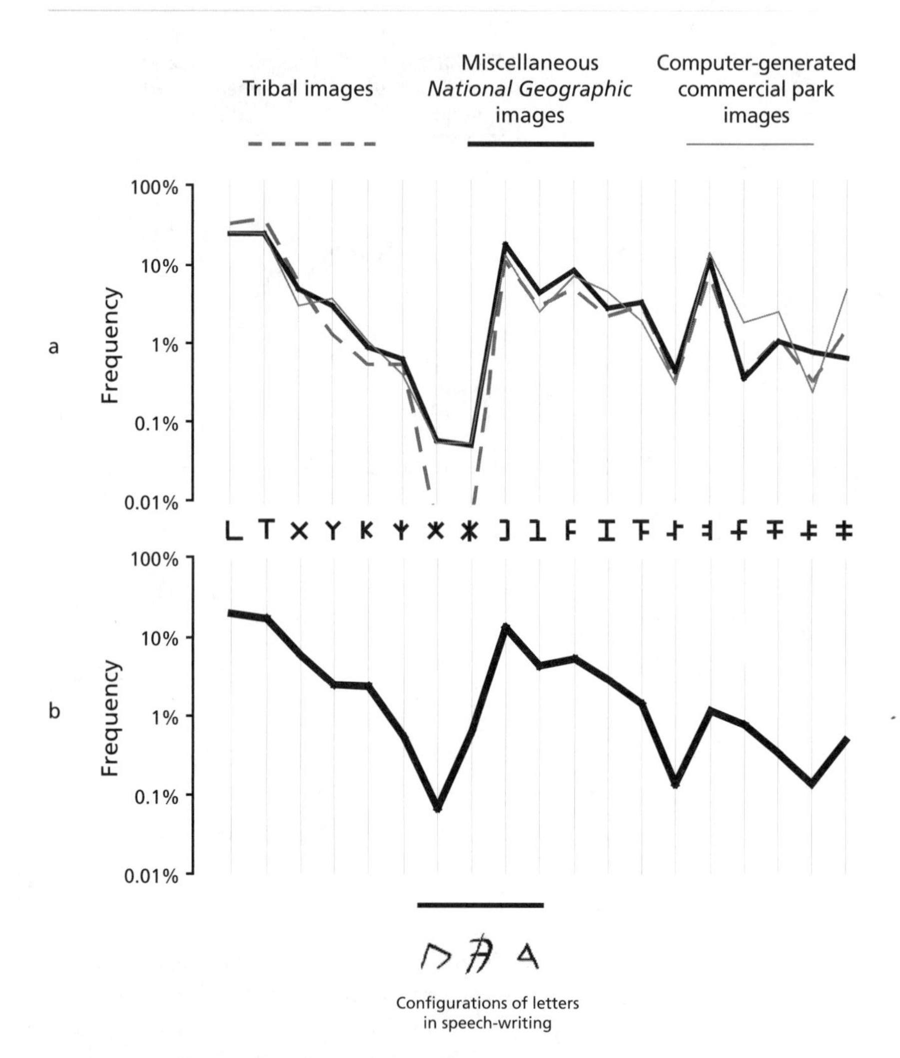

FIGURE 13. *Demonstration that letters have the signature of nature.* (**a**) *The frequency distribution of subthing shapes in nature, as shown in Figure 11 and Figure 12a.* (**b**) *The frequency distribution of shapes within speech-writing. One can easily see how closely the shapes of letters match the configurations found in natural images. Letters in speech-writing systems really are shaped like natural subthings! (Note that this natural signature is not found in random contours, scribbles, or shorthand writing systems.)*

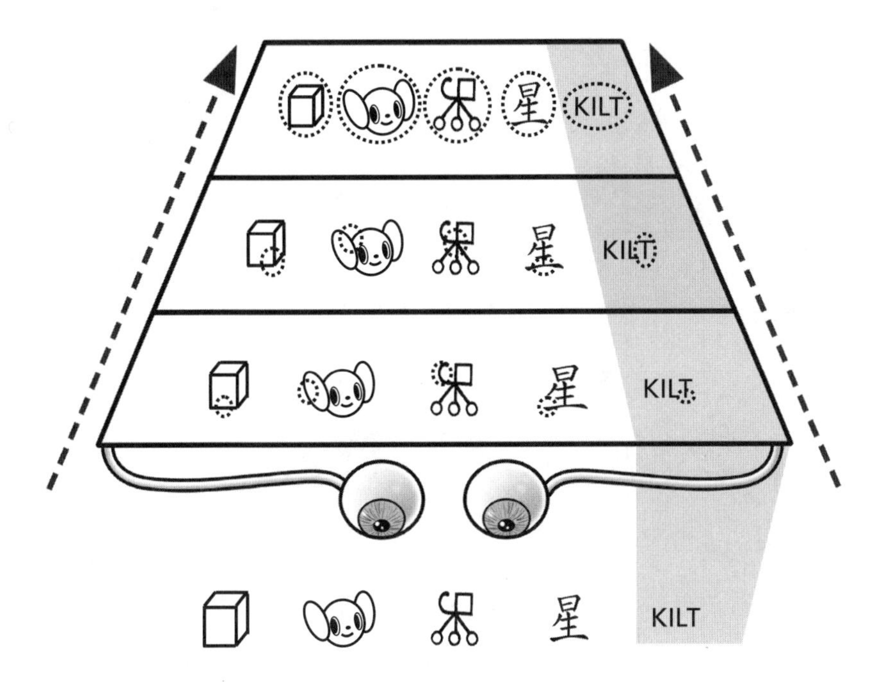

FIGURE 14. *The illustration from Figure 5, modified to include a written word. The illustration shows how the structure of speech-writing nicely harnesses the visual system's object-recognition capabilities by using contour-like strokes that are combined into subthing-shaped letters, which are then combined into object-like words. Neuroscientists Stanislas Dehaene and Laurent Cohen have made a similar suggestion.*

where there are macroscopic opaque objects strewn about, whether in the African savanna, at my new campus at RPI, or on an alien planet. If aliens have macroscopic opaque objects in their midst, and if their culture also developed writing that is good for their eyes, then they have likely ended up with writing that looks similar to the kinds we have here on Earth. In opaque-object environments, L- and T-junctions will tend to be much more common than X-junctions, for example. If our world had instead involved tinted glass objects strewn about, X-junctions would be all over the place, and nature's signature would look quite different.

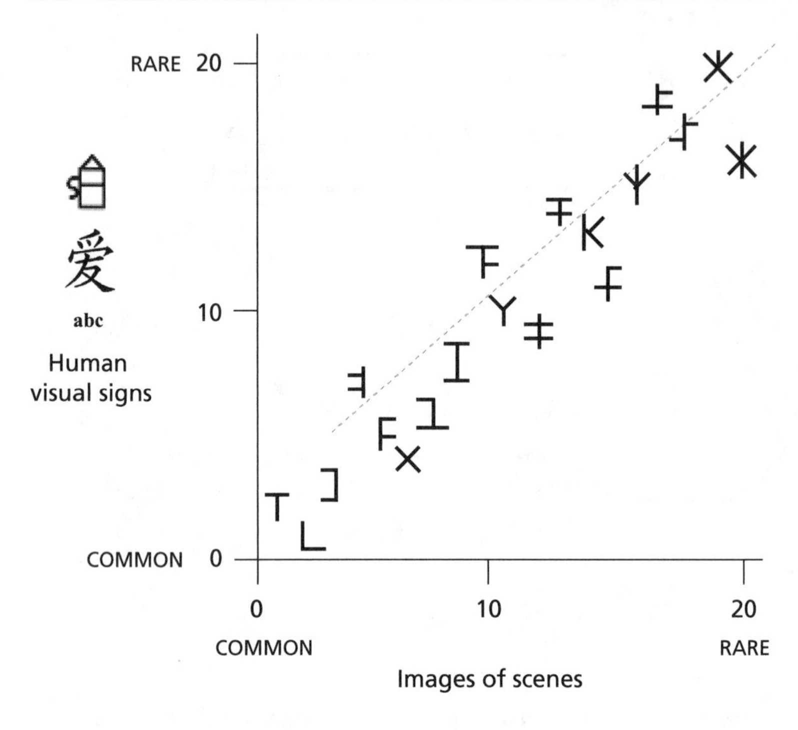

FIGURE 15. *The frequency of configurations in human visual signs and images in nature, plotted in rank order. This shows that the more common configurations in images tend to also be the more common configurations in human visual signs. The visual signs shown in this plot combine non-linguistic signs, Chinese characters, and speech-writing, but the same phenomenon is found in each kind of writing individually, as well.*

Homo turingipithecus

When I was a fellow at Caltech during the fall of 2003, I had a crazy idea. All of us walk around with fantastically powerful computers in our heads. Wouldn't it be great, I thought, if we could tap into that power? The simple act of seeing, for example, relies on the visual brain to complete astoundingly complex computations. Not only does your visual brain carry out countless numbers of these computations per second, but none of these computations *feel* like work. Unlike thinking—in which we grind through ideas in a serial fashion, and which we tend to be conscious of—seeing occurs effortlessly. Furthermore, in vision lots of very complex stimuli can be presented all at once; a picture is worth a thousand inputs. I wondered whether it would be possible to transform software—i.e., programs for processing information—into visual images in such a way that when you looked at these special "software images" your visual brain couldn't help but respond by performing the computations that run the software. The image would be both the software and the inputs the software was designed to process. Your visual system would effortlessly carry out computations—that is, carry them out in a way that would *feel* effortless to you—and generate a perception that told you the solution to the computation. The perception would *be* the output. That is, what I wondered was: Can we trick our visual systems into uploading and computing software of our choosing? This may sound crazy, but it is not any crazier than harnessing DNA to carry out computations, something Leonard Adleman succeeded in making respectable in the 1990s. What's crazier: computations implemented by brains, or computations implemented by vats of DNA soup?

After having my crazy thought, I spent several months trying to create "visual software," and I initially made only meager progress. I even offered $100 and co-authorship to anyone among the Caltech vision community that could find compelling "visual circuits" for computing simple digital circuit components, to no avail. It wasn't until late 2007, while working on this book, that I began to make some real progress.

Figure 16 shows an example "visual circuit." Digital circuits are an important class of computational mechanism used in calculators, computers, phones, and most of today's electronic products; they are built from assemblies of logic gates (functions that carry out simple logical operations like AND, which outputs a 1 only if both inputs are 1), and any part of the circuit—including the output—has a state of either 0 or 1. Unlike electronic circuits, which rely on metal or semiconductors to encode 0 and 1 states, visual circuits like the one in Figure 16 rely on ambiguous visual stimuli—stimuli that can be interpreted two different ways—as their basic components. Every part of the circuit in Figure 16 (except the inputs) can be perceived in two different ways: as tilted toward you or as tilted away from you. Your perception of the tilt is what determines the state of the circuit,

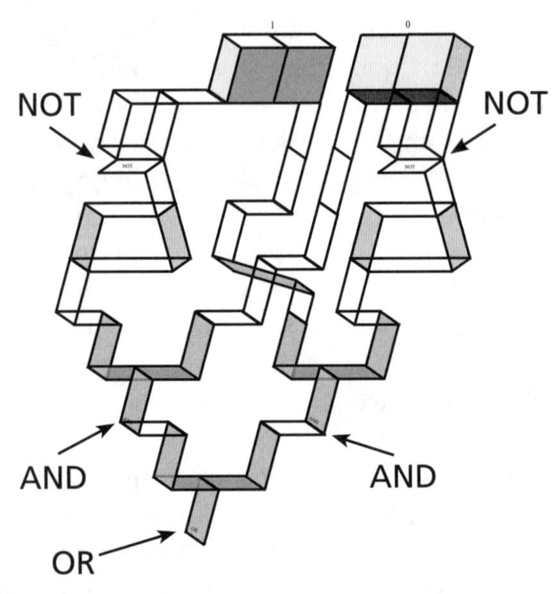

FIGURE 16. *An example of what I call a "visual circuit." The circuit shown here harnesses our visual system to compute an "exclusive OR" (XOR) function (i.e., a function that outputs a 1 when exactly one of the inputs is a 1). The shapes the circuit is built out of are ambiguous stimuli, capable of being perceived in two distinct fashions: tilted toward you, or tilted away from you. The circuit's value, 0 or 1, at any point in the circuit, is determined by your perception. If you perceive a part of the circuit as tilted away from you then the value there is 0; if you perceive it as tilted toward you, then the value is 1. The output here is at the bottom; your perception of the tilt at that point determines the output of the circuit as a whole. This circuit is built out of "wire" (the rectangular lattice), NOT gates (which flip the state of the circuit from tilted toward you to tilted away from you, or vice versa), AND gates (where the output at the bottom of the gate tends to be perceived as tilted toward you only if both inputs above it are perceived as tilting toward you), and an OR gate (the output at the bottom of the gate tends to be perceived as tilting toward you if one or both inputs above it are perceived as tilted toward you). In the case shown, the inputs to this circuit are the shaded cubes at the top, and are unambiguous concerning the direction of tilt at that point in the circuit: they are 1—tilted toward you—and 0—tilted away from you—respectively. This input has a tendency to elicit a perception of "tilted toward you" (1) at the output point, which is expected for an XOR circuit with these inputs. To actually get your visual system to carry out this computation requires that you perceptually "walk through" the circuit from the inputs downward toward the output, although parts of the circuit have a tendency to flip around in an Escher-like frenzy, potentially resulting in a 0 output instead. (The NOT gate variants shown here were inspired by Bram van Heuveln.)*

1 or 0, respectively, at that point in the circuit. The circuit's output is determined by your perception of the tilt at the designated output spot, which in this case is at the bottom. Inputs to a visual circuit are unambiguous stimuli, tilted either toward you (1) or away from you (0). In Figure 16, the inputs are 1 and 0, respectively. Inside the circuit in Figure 16 are three kinds of visual logic gates: NOT gates (which flip the circuit's state from 1 to 0, or vice versa), AND gates (which output 1 only if both inputs are 1), and an OR gate (which outputs 1 if one or both inputs are 1). The AND and OR gates rely on partial visual transparency to help bias one interpretation over the other. For example, the gate at the very bottom of the circuit in Figure 16 is an OR gate, and a casual glance at it will lead to the perception that it is tilted toward you, consistent with the two partially transparent line segments on the gate's left side. The entire circuit in Figure 16 is an exclusive-OR circuit, which means the circuit outputs a 1 when exactly one of the inputs is a 1. Because the inputs here are a 1 and a 0, the output in this case is a 1. That is, because one of the inputs is tilted toward them and one of the inputs is tilted away from them, people will have a tendency to perceive, with some effort, the bottom of the circuit as tilted toward them.

Figure 16 helps explain the fundamental strategy behind visual computation. The visual logic gates in the circuit shown are sufficiently rich that, in principle, any digital circuit (any circuit that works with inputs and outputs of 0 and 1) could be built from them. Of course, larger circuits require smaller and smaller visual circuit components in order to fit them on a single image, and how to do this while still ensuring that the visual system reacts to the circuit as intended is yet another problem, analogous to the difficulties of miniaturization in electronic circuits.

However, before we get overly excited about creating lots of VLSI ("very large scale integrated") visual circuits, I should say that my current version of visual circuits like the one shown has serious difficulties. First, the visual logic gates do not always faithfully transmit the appropriate signal at the output. For example, although AND gates *tend* to elicit perceptions that are AND-like, it is a tendency only, not a surefire physical result as in electronic digital circuits. Second, even if the input unambiguously cues one interpretation, our perception is still somewhat volatile, capable of sudden Escher-like flips to the alternate state. The result is that it can be difficult to actually perceive one's way "through" these visual circuits, although I have experienced considerable personal improvement with practice. The perceived tilt of the output certainly doesn't come effortlessly to the fore with simply a glance at the image, which was my original hope.

As I mentioned above, after my initial failed attempt to trick the visual system into doing computations at my bidding, I temporarily gave up, dejected. Harnessing our visual brain's computational power would have to wait for someone else to solve, I thought. It was not until a year later—during my research into the evolution of writing in 2004—that I began to realize that our visual brains have been harnessed for computation after all! The technology of writing has given us more than just the power to spirit-read and be better listeners. Writing has allowed us to be programmed . . . and not in the *Stepford Wives*, brainwashing fashion. Before writing we were like calculators, able to efficiently carry out a relatively small number of very important kinds of computation. As evolutionary psychologists Leda Cosmides and John Tooby have argued for a couple of decades, we evolved to be good at the tasks that mattered to us, like recognizing a fellow tribesman taking more than his share of meat. We didn't evolve to be general purpose computing machines like our PCs, much less to be programmable by any software engineer passerby.

Although we didn't evolve to be personal computers, could we ever become general-purpose computing machines capable of running arbitrary software? First, let's look at what it would even mean to run software on ourselves. Running software really just means following complex sets of rules, procedures, algorithms, or recipes. Technically, a recipe is a kind of software. When you cook an egg, you follow a simple remembered recipe. You carry out computations, and at each step the computation tells you what to do with the spatula, egg, and pan. But how can a recipe be software—or any kind of computation—when it's not about numbers? Although computer scientists like to think of computations as running solely on numbers, that's only because numbers can stand in for anything. The number 3708 could be the numerical code for "egg." Recipes are still software even if the computations happen to involve vegetables.

If following a recipe counts as running software, then we were able to implement software on ourselves long before the advent of writing. There were, however, severe limits. First, the software had to be remembered by heart. That made software installation very time consuming (memorization takes a lot of rehearsal), and we could only remember so much. We were like the programmable calculators I had in the 1980s and 1990s—programmable in principle, but a pain in the neck because of severe memory limitations. Second, although we humans can, with effort, remember huge amounts of information—think of those people who memorize whole books of the Bible—such huge installations are possible typically because the memorizer has a written text in front of him from which to do the memorizing. Without a written text, memorizers could only learn complex algorithms by hearing them from others. Other people are never as patient as books, even people who have chosen teaching as a profession, and so your teacher would likely break for coffee before you memorized your second verse... not to mention that your teacher would have had to successfully learn the algorithm from someone else in the first place. People did manage to pass on useful knowledge to the next generation, but this tended to require a lifetime's worth of interaction with teachers.

With the advent of writing, we humans went from barely programmable calculators to full-fledged computers capable of computing anything at all. In the first half of the twentieth century, Alan Turing developed a rigorous notion of computation, and a key requirement of his hypothetical Turing machine was that it would have access to an infinite length of tape on which information could be written, stored, and later read from... a tape for reading and writing. The tape was important for a Turing machine because it was an effectively infinite memory space on which to keep track of its calculations. But it was also important for a special kind of Turing machine called a *universal* Turing machine, for such universal machines would know how to read software written onto the tape, and could therefore carry out any algorithm you threw at it.

With the advent of writing we had for the first time our own effectively infinite length of tape. This allowed us to store the contents of our short-term memory onto paper as we carried out our thinking and, more importantly for present purposes, allowed us for the first time to implement any algorithm at all, so long as it had been written down on paper. Rather than having to store the code in our heads, we had the radically simpler task of just following the procedures written on the page, only needing to remember a line at a time in order to implement it before moving on to the next line, and whatever the code told you to do next. Frighteningly complex procedures could be recorded, filling entire books. These "how-to" books related not just lists of facts but complex sets of instructions on how to carry out complex tasks. Amazon.com lists at least 720,000 "how-to" books at the time of this writing, each of which is a piece of software just waiting to be implemented by a human brain.

Our transformation into universal computation machines, however, would have never been possible were writing not so quickly and easily readable. If our recipes were written in binary—i.e., all 0s and 1s—no one would bother trying to read them. By the time you deciphered from the recipe how long the egg should remain on the griddle, the egg would have burned through and turned to ash. If writing poured into the brain like thick ketchup out of a bottle, we might prefer to wait for the teacher to return from his coffee break. He might be less patient than a book, but at least his words slip smoothly and quickly into our brains. The real advance in our personal ability to compute came not with writing per se, but with the kind of writing technology culture developed—writing that harnessed the brain's natural facility for object-recognition. With the invention of writing that is shaped like nature and thereby optimized for the eye, *Homo sapiens* became a qualitatively new kind of animal: *Homo turingipithecus*.

Don't Forget to Write

This chapter has only talked about writing and other simple visual signs. What about the look of fashion, architecture, artifacts, colors, textures, furniture, and, well, everything else in culture?

As a species, we tend to focus on what we *say*, and the giant discipline of linguistics exists specifically to uncover and explain empirical laws governing speech. But our utterances are not just auditory. When we write we are making a visual utterance—and that visual utterance could have evolved to look any number of ways. *All* visual aspects of culture amount to human visual "utterances." But strangely, there has been very little effort to characterize the laws governing them. The only kinds of utterances for which efforts have been made on this front is visual art—but note that the study of auditory art (e.g., music, poetry) hardly counts as serious linguistics, and accounts for at most a tiny fraction of the field. Just as linguistics complements laboratory studies in helping us understand the brain and cognition, we should expect that studying the visual utterances people make—studying what I call *visual linguistics*—would aid in the comprehension of our visual systems. This field of visual linguistics is essentially a missing discipline, and I consider this chapter to be research within this largely unexplored discipline (see Figure 17).

We started this chapter with the superpower of spirit-reading, a power we sometimes fantasize about, but a power that only the deluded tend to believe they actually possess. Yet we *do* have this power, and writing is what makes it possible. But writing only gives us this power because it has been designed by cultural evolution to be easily understood, and culture was only able to achieve this by using shapes that fit easily into the brain—namely, the shapes of nature. We saw that when a written language *can* have object-like symbols for words—when a language is logographic—it tends to have them. And when writing can't—like in speech-writing, where symbols stand for speech sounds—culture *still* gives it the old college try, making letters look like natural object *parts*.

Although this technological innovation by culture gave us the spirit-reading superpower, spirit-reading is hardly why writing is so super, and probably has relatively little to do with why writing itself was selected for. In all likelihood this book will be read by others only while I'm alive. If you're reading this before 2069 (the year I turn 100), then I hope for my sake you're not spirit-reading. But in about sixty years, when I'm finally dead and gone, I suspect few people will still be reading this as their source for the material I discuss here. Although I'd be delighted if this book were sitting on the backs

	Laboratory experiments	Human "utterances"
Cognition	Cognitive psychology, cognitive neuroscience	Linguistics
Vision	Visual psychophysics	? Visual Linguistics

FIGURE 17. *There ought to be a discipline that plays the same role for vision that linguistics does for cognition. That is, there should be a discipline that studies the laws governing the natural visual "utterances" people make—writing, visual signs, fashion, architecture, and the other visual aspects of culture—and relates these empirical regularities to the visual system and its evolution.*

of twenty-second-century toilets all over the world, I'm writing this book in order to convey ideas from the breathing to the breathing. I'm putting knowledge, ideas—and a dollop of entertainment—onto paper. And that is presumably the central power writing gave us, and the one that has affected the trajectory of human culture most deeply: the ability to effectively communicate our ideas to others.

Also more important than spirit-reading is the radical increase in computational power writing gave us, turning us from *Homo sapiens* to *Homo turingipithecus*. Efficient writing allowed us to offload our in-the-head calculations and ponderings onto the backs of envelopes. And it enabled us for the first time to install "how-to" software into ourselves. Reading a *How to Pilot a Helicopter* book takes a lot longer than Trinity in the movie *The Matrix* needed to download the same information into her brain, but from our preliterate ancestors' point of view, they'd be amazed that you can literally hold that kind of information in your hand; a book is essentially just a bulky floppy disk.

Not only are the recording of knowledge and personal computer upgrades more important than spirit-reading, but they are what make spirit-reading worthwhile in the first place. With all due respect to my preliterate ancestors, without the vast access to facts and ideas and the computational power that the invention of writing allows, they probably wouldn't have had much writing worth spirit-reading.

The world has shaped our eyes, something true of all animals, for the world has seen to it that the owners of poorly performing eyes tend to find it difficult to reproduce. Natural selection is the major source of our visual superpowers, and was central to three of my four supervision stories: color vision from chapter 1 was selected for so that we might see emotions and other states on the skin; forward-facing eyes from chapter 2 were selected for so we could use X-ray vision in cluttered environments; and the illusions in chapter 3 were a consequence of the future-seeing power selected for so that we might perceive the present.

Natural selection shaped human eyes just as much as it did every other animal's, but our eyes have an additional distinction: our eyes reach out and shape the world! They—and we—do this through culture. This means that there is a second way to evolve a visual superpower, one that is crucial to understanding the story on spirit-reading in this final chapter: we can, through culture, change the world to fit the eye. Natural selection made the eye good at processing objects in nature, and so culture evolved visual signs that had properties akin to those of objects in nature, so that the eye could process them as optimally as possible.

The common denominator in our superpowers of vision is that they are all the result of evolution, whether through natural selection or cultural selection, and so only by examining vision in its evolutionary context can we hope to make landmark discoveries on the nature of our powers: there's no revolution without evolution.

Index

L

M

N

O

P

R

S

About the Author

MARK CHANGIZI carried out most of the research in the book as a Sloan-Swartz Fellow in Theoretical Neurobiology at the California Institute of Technology, and is now an assistant professor in the Department of Cognitive Science at Rensselaer Polytechnic Institute. His research areas tend to concern the evolutionary function and design principles governing complex behaviors, perceptions, and organisms. His first book appeared in 2003 and is called *The Brain from 25,000 Feet: High Level Explorations of Brain Complexity, Perception, Induction and Vagueness* (Kluwer Academic, Dordrecht). Dr. Changizi is the first author on more than thirty journal articles in diverse topics, and his research has been covered in more than one hundred and fifty media outlets worldwide, including *The New York Times*, *Time*, *Newsweek*, *USA Today*, *Financial Times*, *Daily Telegraph*, *The Times of London*, *Scientific American*, *Discover*, *New Scientist*, *Natural History*, *Wired*, Reuters, ABC News, MSNBC, and Fox News.